1 MONTH OF
FREE
READING

at

www.ForgottenBooks.com

By purchasing this book you are eligible for one month membership to ForgottenBooks.com, giving you unlimited access to our entire collection of over 1,000,000 titles via our web site and mobile apps.

To claim your free month visit: www.forgottenbooks.com/free12480

ISBN 978-0-365-05076-6
PIBN 10012480

PHYSICS

FOR

UNIVERSITY STUDENTS

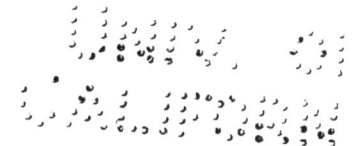

BY

HENRY S. CARHART, LL.D.

PROFESSOR OF PHYSICS IN THE UNIVERSITY OF MICHIGAN

PART I.

MECHANICS, SOUND, AND LIGHT

REVISED EDITION

𝔅𝔬𝔰𝔱𝔬𝔫

ALLYN AND BACON

1898

CAJORI

PRESS OF
Rockwell and Churchill
BOSTON, U.S.A.

PREFACE.

THE prevailing practice in giving instruction in Physics to undergraduate students in American colleges and universities is a judicious combination of the text-book and the lecture systems. This is particularly true for a first course covering the entire subject, and laying a foundation of general principles for more advanced study by special courses and laboratory work. This practice the author has followed in teaching large classes for many years; but, finding no book which meets his needs, he has felt impelled to prepare his own text. The present volume is an extension of one written several years since for the sole use of his classes. Fellow teachers into whose hands the earlier book has fallen have encouragingly advised the expansion of it into a volume of somewhat less modest pretensions, for the benefit of others who employ similar methods. The result is a text-book and not a treatise on Physics. No attempt has been made to secure completeness. The book is not a cyclopedia of Physics. Only such topics have been selected as appear most important from the point of view of a general survey of the science; and an effort has been made to place them in a logical relation to one another. Somewhat more attention has been given to Simple Harmonic Motion than is customary in an elementary course. Its extensive application in the study of alternating currents of electricity, added to its earlier

911249

important service in Mechanics, Sound, and Light, renders a more thorough study of it imperative.

The book is not intended to take the place of the living teacher. It leaves room for the personal equation in instruction ; but it will relieve the student of a large part of the labor of taking notes; and, it is hoped, will secure for him more accurate statements than he would be likely to obtain from listening to lectures without the aid afforded by a text-book of principles.

In many cases the method of the calculus has been employed without its formal symbols. The course in Physics represented by this book is supposed to precede the study of the calculus, and the methods used will prepare the student for the employment of that branch of mathematics in more advanced courses

Free use has been made of the books referred to in the headings of articles, and especially of Violle's admirable *Cours de Physique.*

The present volume covers the work done in the first course extending over one-half of the academic year. The second part will be devoted to Heat, Electricity, and Magnetism.

The author's thanks are due to Assistant Professor J. O. Reed for many valuable suggestions and for careful reading of the proof sheets.

A few of the cuts in Sound and Light were kindly furnished by the publishers of Anthony and Brackett's Physics, with the permission of the authors.

University of Michigan, September, 1894.

CONTENTS.

MECHANICS.

REFERENCES.

The letters, enclosed in brackets accompanying the headings of articles, refer to the following books, numerals denoting pages:

A. and B., Anthony and Brackett's *Text-Book of Physics.*

B., Barker's *Physics.*

Bl., Blaserna's *Theory of Sound.*

D., Daniell's *Text-Book of the Principles of Physics* (Second Edition).

H., Helmholtz's *The Sensations of Tone*, translated by Ellis; the second numerals in small brackets, the German text of *Die Lehre von den Tonempfindungen.*

K., Koenig's *Quelques Expériences d'Acoustique.*

L., (in Mechanics) Lodge's *Elementary Mechanics.*

L., (in Light) Lommel's *The Nature of Light.*

M. and M., Maxwell's *Matter and Motion.*

P., Preston's *Theory of Light.*

S., Spottiswoode's *Polarization of Light.*

T. and T., Thomson and Tait's *Elements of Natural Philosophy.*

T., Tait's *Light.*

Tyn., Tyndall's *Sound.*

V., Violle's *Cours de Physique.*

Z., Zahm's *Sound and Music.*

MECHANICS.

CHAPTER I.

INTRODUCTION.

1. Physics Defined. — Physics formerly comprised the study of all the phenomena of nature. In more recent times it has abandoned the history of organized life on the one hand, and the study of celestial phenomena on the other; and it is now restricted to the study of the general properties of matter, particularly to those phenomena which involve changes in the energy associated with matter, without altering the essential constitution of bodies.

The real nature of matter is unknown to us. We know something about its properties, that it is conserved in quantity, that it apparently occupies space and affects our senses. It is preëminently the vehicle of Energy, the capability which one body possesses of producing motion or effecting changes in another body.

Modern Physics is the exposition of those phenomena of nature which involve corresponding changes in the associated energy. Formerly it was largely a study of matter; now it is largely a study of energy.

2. Boundaries of Physics. — It must not be assumed that Physics is separated from related sciences by sharply defined boundaries. It overleaps the " metes and bounds " set for it in scientific classification and enters the domain of its nearest neighbors. Since all natural phenomena involve energy, and Physics is largely a study of energy, it is evident that every natural science furnishes abundant material for physical investigation. A science is classified in accordance with the chief aims to which it is devoted. Thus Chemistry studies the atomic organization and structure of the molecule. It endeavors to ascertain how the so-called elements enter into the composition of both organic and inorganic bodies, and the laws governing all those inner changes in matter which affect its properties and identity. The combustion of carbon, the rusting of iron, the burning of limestone, the fermentation of wine, are changes involving the constitution of the molecules of the several bodies. They are therefore chemical changes. At the same time they involve energy changes, and are therefore appropriate studies for the physicist. In fact, there are so many problems in which Physics and Chemistry are equally interested that a new subdivision of science, called Physical Chemistry, is even now in process of differentiation. So also a large part of modern astronomy is so much concerned with the physical constitution of celestial objects, with their intrinsic and variable brightness, with the self-luminosity of suns and the panorama of nebulous condensation, that it has won for itself the title of Physical Astronomy, or Astro-Physics.

Mental Philosophy and Natural Philosophy, or Physics, were once both characterized by the employment of the metaphysical method. The adoption of the inductive method and the experimental plan of attack in Physics led this branch of study far away from its ancient ally; but now Psychology has appropriated the methods of Physics, and the two are moving along the converging lines of experimental research.

But energy may be transferred from one body to another, and may be transformed into the various forms which it is capable of assuming, without any changes in the constitution of matter. .Physics has, therefore, a province independent of chemistry, physiology, and other allied branches.

3. The Method of Physics. — A science is characterized no less by the method essential to it than by its aim. The aim of Physics is to ascertain the causal connection between related phenomena. Such study is justified by the accumulated experience of the human race, that under the same conditions the same results flow from the same causes. In other words, the events of nature are not fortuitous. This principle is embodied in the expression, *the constancy of the order of nature.*

But to ascertain the causes of physical phenomena and the laws of action of physical forces, it is absolutely essential that the consequences of any provisional theory of them should be subjected one by one to the control of experimentation. In experiment the phenomena are produced under conditions controlled by the operator. In this way the necessary relationships are established, and the results flowing from a single cause are separated from all others.

Causal relations suggested by an attentive observation of phenomena, and supported by reasoning and experiment, must in the end be subjected to the ultimate test of *measurement.* Modern Physics is essentially quantitative in character. It is not enough to know that a relation exists, but that relation must be expressed numerically.

4. Fundamental Units of Measurement. — Since all the phenomena of nature occur in *matter*, and are presented to us in *time* and *space* relations, physical measurements involve the choosing of three fundamental units, viz., the unit of *length*, the unit of *mass*, and the unit of *time*. This particular selection is, however, a matter of convenience rather than necessity, and rests upon several considerations which properly determine the choice of these fundamental quantities.

All other units employed in Physics are defined in terms of those of length, mass, and time. They are, therefore, called *derived* units, in distinction from the other three, which are called *fundamental* units.

The three fundamental units now universally employed in Physics are the *centimetre*, the *gramme*, and the *second ;* and the system of measurement founded upon them is called the C.G.S. or absolute system.

5. The Unit of Length. — The *centimetre* is the hundredth part of the length of a bar of platinum at 0° C., preserved in the national archives at Paris, and known as the *mètre des archives.* It was constructed in accordance with a decree of the French Republic passed in 1795 on recommendation of a committee of the Academy of Sciences.

The value of the metre in the earlier feet of different countries is as follows:

	Foot in Metres.	Metre in Feet.
France	0.3248394	3.078444
Austria	0.3161109	3.163446
Prussia and Denmark	0.3138535	3.186199
England and Russia	0.3047945	3.280899
Baden and Switzerland	0.3000000	3.333333

	Foot in Metres.	Metre in Feet.
Sweden	0.2969010	3.368126
Hanover	0.2920947	3.423547
Bavaria	0.2918592	3.426310
Hesse	0.2876991	3.475854
Würtemberg	0.2864903	3.490519
Saxony	0.2831901	3.531197

By Act of Congress of the United States in 1866 the metre was defined to be 39.37 inches.

The metre was intended to be the ten-millionth part of an earth-quadrant from the equator to the pole. It is now known that such a quadrant is about 10,002,015 metres. This difference between the *ideal* and the *legal* metre illustrates the difference between a *theoretical* and a *practical* **unit.**

6. The Unit of Mass. — Mass is the quantity of matter in a body. The unit of mass in the C.G.S. system is the *gramme.* Theoretically it is the mass of a cubic centimetre of distilled water at the temperature of maximum density, or 4° C. Practically it is the 1-1000th part of a standard mass of platinum preserved in the archives at Paris and called the *kilogramme des archives.* The theoretical and practical definitions again turn out to be not absolutely identical. National prototypes of the metre and the kilogramme, made by an International Commission, are preserved in the Bureau of Weights and Measures in Washington.

The gramme was selected as the unit of mass because of its convenience, since it is nearly the mass of a unit volume of water at maximum density; and as water is taken as the standard in determining specific gravity, it follows that density, or the mass of matter in a unit volume, and specific gravity are numerically equal.

7. The Unit of Time. — The unit of time universally employed is the *second* of mean solar time. An apparent solar day is the interval between two successive transits of the sun's centre across the meridian of any place. But the apparent solar day varies in length from day to day throughout the year by reason of the varying speed of the earth in its orbit. Hence the mean or average length of all the apparent solar days throughout the year is taken. This is divided into 86,400 equal parts, each of which is a second of mean solar time.

8. General Subdivisions of the Subject. — It is convenient to make two general divisions of the subject-matter of Physics:

1. *Physics of Matter.* This includes the laws of Motion, the mechanics of Solids and Fluids, and Sound.

2. *Physics of the Ether.* The ether may be a refined kind of matter. At all events, it is desirable to consider by themselves those branches which deal particularly with the ether, and which are not completely explicable without taking this medium into account. They are Heat, Light, Electricity, and Magnetism.

That branch of the first general subdivision which investigates the action of *force* is called by most logical writers *Dynamics*. It is commonly called *Mechanics*, a term employed by Newton to designate the science of machines and the art of making them.

Since force is known only by the motion it produces, a discussion of the laws of pure motion should precede that of the laws of force. This constitutes the subject-matter of *Kinematics*. To the idea of space involved in geometry is added in this branch that of time.

Dynamics is divided in accordance with the two methods in which force is recognized as acting, viz. :

a. As preventing motion or change of motion.

b. As producing motion or change of motion.

Dynamics, therefore, includes *Statics*, in which equilibrium under the conjoint action of two or more forces is considered ; and

Kinetics, in which the relation of forces to motion is studied.

Our chief attention will be given to Kinematics and Kinetics.

CHAPTER II.

KINEMATICS.

9. Motion (M. and M., 38). — The configuration of a system of points or minute material particles is their relative positions. When a change of configuration is considered only with respect to its state at the beginning and end of the process of change, without respect to time, it is called *displacement.*

But when the attention is directed to the process itself, as taking place within a certain period and in a continuous manner, then the change of configuration is called *motion.*

Motion is the change in the relative position of a material particle.

A *material particle* is supposed to be without dimensions, but to possess all the properties of matter.

Kinematics is the science of motion considered in itself and apart from the causes producing it. Ampère separated it from rational mechanics as a branch by itself, and gave to it the name which it now bears.

It is of consequence to observe that all motion is merely relative, since there are no fixed points in space to which absolute motion can be referred. Maxwell says: "Any one who will try to imagine the state of a mind conscious of knowing the absolute position of a point will ever after be content with our relative knowledge."

10. The Path. — The succession of positions of a mate-rial particle is called its path or trajectory. The path of a material particle must always be continuous. There can be no abrupt change of velocity or of the direction of mo-tion; such abrupt change would imply the action of an in-finite force. By abrupt change is meant one occurring in zero time. Two consecutive portions of the path of a point would then come together at an angle. Such a change would require an infinite force. But if the angle is rounded off with a small arc, then time is involved in the change and a finite force may effect it.

If any point of a path be selected, the particle will pass through that point at least once during the motion. It may pass through it more than once if the path is curvilinear.

Mathematical curves may be discontinuous, but a ma-terial particle cannot traverse a discontinuous path while retaining continuous existence in time and space.

11. Direction of Motion and Curvature (T. and T., 2). — The direction of motion at any point of a curved path is the geometrical tangent to the curve at the point. The direction of the motion then changes from point to point along the curve. The rate of change of direction per unit length of the curve is called the *curvature*. This may be either constant or variable. It is constant in a circle or a helix; it is variable in an ellipse or a parabola. The cur-vature of a straight line is zero.

The curvature at any point of a curve is the reciprocal of the radius of the circle which most nearly coincides with the curve at the point. This may be demonstrated as follows:

Consider the curvature between two contiguous points *P* and *Q* (Fig. 1). Let θ be the angle between the two radii *PO* and *QO*, and let *s* be the arc *PQ*. Then θ is also the angle between the tangents at *P* and *Q*, and this angle is the entire change in direction in passing from one point to the other. The curvature is therefore θ/s. Let *r* be the radius. Then

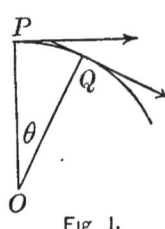

Fig 1.

$$r\theta = s,$$

and

$$\frac{\theta}{s} = \frac{1}{r}. \quad \ldots \ldots \quad (1)$$

If the two points are indefinitely near together, *r* is the radius of the osculating circle. The curvature is therefore equal to the reciprocal of the radius of this osculating circle.

In the case of a circle the angular change in direction in going once round is 2π. The curvature is this angle divided by the arc traversed, or by $2\pi r$. Therefore the curvature for a circle is

$$\frac{2\pi}{2\pi r} = \frac{1}{r}.$$

12. Speed and Velocity.—Time-rate of motion, without reference to the direction of motion, is called *speed*. Thus we speak of the speed of a horse, the speed of a cannon ball, the speed of a railway train.

But when the motion is a directed quantity, that is, along a definite line whose direction is given, the rate of motion is then called *velocity*.

When the material point traverses equal spaces in equal times along a right line, it describes *uniform rectilinear*

motion. The velocity along the line is the constant space
traversed in the unit of time.

$$O \quad A \qquad M \qquad X$$

Fig. 2

Let the distance $OM = s$
(Fig. 2) be described in the
time t, this distance being measured along the path OX
from the point O, where the point is at the time $t = 0$.

Then $\qquad\qquad\qquad s = vt.$

From which $\qquad\qquad v = \dfrac{s}{t},\quad .\ .\ .\ .\ .\ .\ (2)$

or the velocity is the constant ratio of the space to the
time employed in describing it. It may be represented
geometrically by the path OA traversed in the unit of time.

Both the space s and the time t may be reduced to
infinitesimal values without affecting their ratio. Call
the infinitesimal space ds and the corresponding time required to describe it dt; then

$$v = \frac{ds}{dt}.$$

If the motion is *rectilinear but not uniform*, let M and M'
(Fig. 3) be the positions of
the material particle at the
two neighboring epochs t and

$$O \qquad M \quad M' \qquad A \quad X$$

Fig. 3.

t'. Then $v_1 = \dfrac{MM'}{t'-t}$ is the mean velocity during the time
$t'-t$. It is the constant velocity with which a particle,
travelling along the right line OX, would describe the
space MM' in the time-interval $t'-t$. But it is not the instantaneous velocity at the time t. Imagine now MM' to be
reduced to infinitesimal dimensions, that is, to tend toward
a value of zero; and let its value then be represented by ds
as before. Also let the time of traversing this minute
distance be dt. Then precisely as before

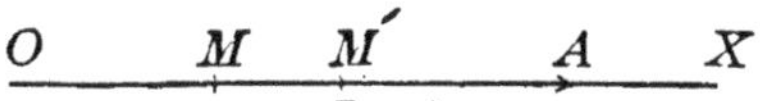

$$v = \frac{ds}{dt}.$$

Geometrically this velocity may be represented by the line MA laid off from M toward X. The velocity at the time t is the distance which the body would describe if it should continue to move *uniformly* from that instant for a unit of time.

The practical unit of velocity is the velocity of *one centimetre per second*.

13. Acceleration. — When a particle traverses a right line with variable velocity, a case of special interest presents itself where equal changes of velocity take place in equal intervals of time. The motion is then uniformly accelerated (or retarded), and the constant change of velocity in the unit of time is called the *acceleration*. *Acceleration is the time-rate of change of velocity*. Let v_0 be the initial velocity at time 0, and let v be the final velocity at time t; also let a be the acceleration. Then from the definition

$$a = \frac{v - v_0}{t}; \quad \ldots \ldots \ldots \quad (3)$$

whence
$$v = v_0 + at. \quad \ldots \ldots \ldots \quad (4)$$

If the time t is reduced to the infinitesimal value dt and the corresponding velocity-change is dv, then

$$a = \frac{dv}{dt}.$$

This expression is called the derivative of the velocity with respect to the time. It applies equally well to either uniform or variable acceleration.

The practical unit of acceleration is an acceleration of one centimetre per second per second; that is, it is a change in velocity of one centimetre per second taking place in a second of time.

Acceleration may be either positive or negative. A negative acceleration is a retardation.

PROBLEMS.

1. If a body start from rest with a uniform acceleration of 2 metres per second, find its velocity at the end of 3 minutes.

2. A body starts with a velocity of 300 metres per second. If it comes to rest in 1 minute 2.5 seconds, find the uniform negative acceleration.

3. A body has an initial velocity of 6 metres per second; find its velocity at the end of 1, 2, 3, and 6 seconds respectively, if a equals 9.8 m.

14. Space described in Uniformly Accelerated Motion. — In uniformly accelerated motion the equivalent velocity, or the velocity with which a uniformly moving particle would describe the same space in the same time, is equal to the half sum of the initial and final velocities. Hence

$$s = \frac{v_0 + v}{2} t. \quad \ldots \ldots \quad (5)$$

Geometrically this may be represented by means of a right triangle (Fig. 4), in which the vertical lines drawn at small equal distances represent the velocities at successive instants of time. These velocities form an arithmetical progression, since the variation is constant. Let the equal divisions along the base of the triangle be the

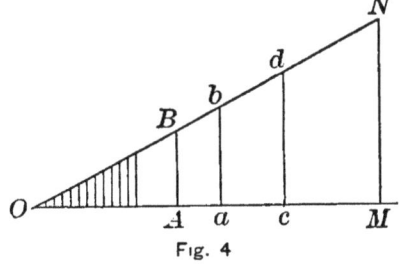

Fig. 4

small time-interval dt. Then if the particle had started from rest at O, the mean velocity would be represented by the line ab midway between O and M, and this equals half the final velocity MN. But if the initial velocity is

AB, then the mean equivalent velocity is *cd,* the line drawn midway between *A* and *M.* But *cd* is equal to the half sum of *AB* and *MN,* or the mean velocity is

$$\frac{v_0 + v}{2}.$$

Moreover, each instantaneous velocity $\frac{ds}{dt}$ multiplied by the small time-interval *dt,* during which the velocity may be assumed to be constant, is *ds,* the small space described during time *dt.* This space is one of the narrow strips making up the triangle. Hence the entire space described during the time *t* is numerically equal to the area of the figure *AMNB,* which is the expression in equation (5).

If we substitute in equation (5) the value of *v* obtained from equation (4), we have

$$s = \frac{2v_0 + at}{2} t = v_0 t + \tfrac{1}{2}at^2. \quad . \quad . \quad . \quad (6)$$

Multiply together (3) and (5) and

$$as = \frac{v^2 - v_0^2}{2},$$

or
$$v^2 = v_0^2 + 2\,as. \quad . \quad . \quad . \quad . \quad (7)$$

If the initial velocity is zero, then (4), (6), and (7) become

$$v = at \quad . \quad . \quad . \quad . \quad . \quad (8)$$
$$s = \tfrac{1}{2}at^2 \quad . \quad . \quad . \quad . \quad . \quad (9)$$
$$v^2 = 2as. \quad . \quad . \quad . \quad . \quad (10)$$

When *t* is unity (9) becomes

$$s = \tfrac{1}{2}a,$$

or the space traversed in unit time is half the acceleration. If *t* is unity in (8) the velocity equals the acceleration. Therefore the space described during the first second, when the body starts from rest, is half the instantaneous velocity at the end of the second.

15. Second Method. — Let the whole time t be divided into a very large number n of equal time-intervals τ.

Then $n\tau = t$.

If the particle starts from rest, the velocities at the end of the several small time-intervals are

$$a\tau, \; 2a\tau, \; 3a\tau, \; . \; . \; . \; na\tau.$$

If these time-intervals are sufficiently small the velocity during each interval may be assumed constant. Then the spaces described in the successive intervals are

$$a\tau^2, \; 2a\tau^2, \; 3a\tau^2, \; . \; . \; . \; na\tau^2.$$

The entire space is the sum of these elementary spaces, or

$$s = a\tau^2 \, (1 + 2 + 3 + \; . \; . \; . \; n).$$

To find the sum of the series in the parenthesis apply the following theorem: When n is *indefinitely large* the sum of the mth powers of the natural numbers **1, 2, 3,** etc., to n is $\dfrac{n^{m+1}}{m+1}$, where of course m is the exponent.[1]

---

[1] To find the sum of the series

$$1^m + 2^m + 3^m + \; . \; . \; . \; n^m$$

we have (Todhunter's Algebra, p. 404, Hall and Knight's Algebra, p. 336).

$$s = Cn^{m+1} + A_0 n^m + \frac{m}{2} \, A_1 n^{m-1} + \frac{m(m-1)}{2 \cdot 3} \, A_2 n^{m-2} + \text{etc., where}$$

$$C = \frac{1}{m+1}; \quad A_0 = \tfrac{1}{2}; \quad A_1 = \tfrac{1}{6}; \quad A_2 = 0.$$

All coefficients with even subscripts are zero. Hence

$$s = \frac{n^{m+1}}{m+1} + \tfrac{1}{2}n^m + \frac{m}{12} \, n^{m-1} - \frac{m \, (m-1) \, (m-2)}{2 \cdot 3 \cdot 4 \cdot 30} \, n^{m-3} \text{ etc.}$$

If we suppose the series cleared of fractions,

$$as = n^{m+1} + bn^m + cn^{m-1} - \text{etc.,}$$

a series of descending powers of n.

Omitting constants, s is of the form

$$s = n^{m+1} + n^m + n^{m-1} - \text{etc.}$$

But $\quad n^{m+1} = n^m \times n$, and $n^{m+1} + n^m = n^m (n+1)$.

Therefore $\qquad 1 + 2 + 3 + \ldots n = \dfrac{n^2}{2},$

and $\qquad\qquad s = a\tau^2 \dfrac{n^2}{2} = \tfrac{1}{2}at^2,$

since $\qquad\qquad n^2\tau^2 = t^2.$

PROBLEMS.

1. If a body with uniform acceleration acquire a velocity of 10 metres a second in moving a distance of 25 metres from rest, find the acceleration.

2. In what time will a body moving with a uniform acceleration of 9 metres a second traverse 250 metres?

3. A cannon ball has a muzzle velocity of 400 metres a second; the length of the gun traversed by the ball is 3 metres. Find (a) the acceleration on the assumption that it is uniform; (b) the time of traversing the gun.

16. The Free Fall of Bodies. — The general formulas connecting space, time, velocity, and acceleration in uniformly accelerated motion have already been developed in Articles 13 and 14. If we consider the acceleration of gravity constant at any place on the earth's surface, and call it g, then we obtain the following formulas by substituting for a the particular acceleration in question, g. Hence

$$\left.\begin{array}{l} v = gt \\ s = \tfrac{1}{2}gt^2 \\ v^2 = 2gs. \end{array}\right\} \quad \ldots \ldots (11)$$

If now n is *indefinitely large*, then unity may be omitted in comparison with it, and we have for the sum of the first two terms of the series

$$n^{m+1} + n^m = n^m \times n = n^{m+1}.$$

Under the above conditions, therefore, the sum of the first two terms equals the first term, or the second term may be omitted in comparison with the first. The succeeding terms are all still lower powers of n, and they are therefore negligible in comparison with the first term. It follows finally, then, that

$$s = \frac{n^{m+1}}{m+1}.$$

This formula will be found to be very useful in several subsequent topics.

If the body starts with an initial velocity v_0, then

$$\left.\begin{array}{c} v = v_0 \pm gt \\ v^2 = v_0^2 \pm 2gs \\ s = v_0 t \pm \tfrac{1}{2}gt^2. \end{array}\right\} \quad \ldots \quad (12)$$

The plus sign is applicable to motion downward and the minus sign to projection upward.

PROBLEMS.

1. A particle has a uniform acceleration of 20 cms. a second, and an initial velocity of 30 cms. a second. Find (*a*) the velocity after 16 seconds; (*b*) the time required to travel 300 cms.; (*c*) the change in velocity in traversing that distance.

2. A body is projected upward with any velocity, and t and t' denote the times during which it is respectively above and below the middle point of its path; find the value of $\dfrac{t}{t'}$.

3. A body dropped from the top of a tower 20 metres high reached the bottom of a well within the tower in 3 seconds; find the depth of the well. The acceleration of gravity is 9 8 metres per second.

17. Projection Upward. — If motion upward is considered positive, then the acceleration is negative and equations (12) must be applied with the minus sign. The following three problems may then be solved:

(*a*) *To find the time of ascent.* When the particle reaches the highest point, $v = 0$. Hence

$$v_0 = gt \quad \text{and} \quad t = \frac{v_0}{g}.$$

(*b*) *To find the height to which the particle will ascend.* Since at the highest point $v = 0$,

$$v_0^2 = 2gs \quad \text{and} \quad s = \frac{v_0^2}{2g}.$$

But this is the velocity which a body acquires in falling

freely through a height *s.* Therefore, neglecting atmospheric resistance, that is, allowing the particle to move
freely, it will return to the point of projection with the initial velocity in the opposite direction.

(*c*) *To find the time when the particle will be at a given
height.* Substitute the given height for *s* in the third
equation. This gives an equation of the second degree
with two roots. The physical interpretation is that the .
particle will be at the given elevation once while ascending and again when descending.

PROBLEMS.

1. A body is projected vertically upward with an initial velocity
of 250 metres per second. Find (*a*) how high it will rise; (*b*) the
time of ascent; (*c*) when it will be 350 metres above the starting
point.

2. A body thrown vertically upward passes a point 10 metres
from the starting point with a velocity of 20 metres a second. Find
(*a*) how much farther it will go; (*b*) its initial velocity.

**18. Composition of Motions (A. and B., 16; T. and T., 8;
V., I, 48; B., 30).** — Motions, velocities, and accelerations
are directed quantities, and they are capable of geometrical
addition. When a particle has several motions given to
it at the same time, its actual motion is made up of all of
them. The motions are then said to be compounded.
The actual motion is the *resultant*, and the several motions
entering into it are called the *components*.

When there are two uniform rectilinear motions along
the same straight line, their resultant is their sum or
difference, according as they are similarly or oppositely
directed.

Let us now consider the composition of two uniform rectilinear motions taking place along different right lines. Let the particle be subject to a uniform motion along the line OX (Fig. 5), while at the same time this line is displaced parallel to itself, so that the point O moves uniformly along OY.

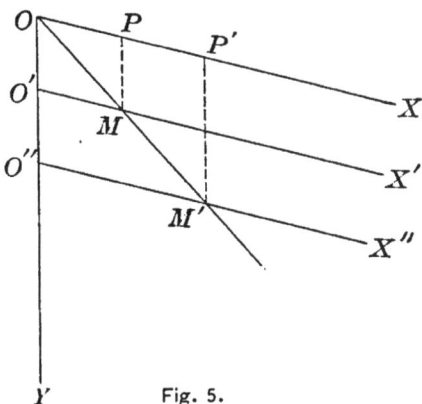

Fig. 5.

In the time t the particle moving along OX arrives at the point P, and

$$OP = vt.$$

But OX is displaced to the parallel position $O'X'$, and

$$OO' = v't.$$

The particle is then at M, the line $O'M$ being taken equal to OP. PM is therefore parallel to OO'.

In another time t' the particle will arrive at P' along OX, and

$$OP' = vt'.$$

In the same time the line will have arrived at the position $O''X''$ and

$$OO'' = v't'.$$

The particle will then be at M', the point of intersection of the parallel to OY through P' with $O''X''$.

But

$$\frac{OP}{OP'} = \frac{t}{t'} = \frac{OO'}{OO''} = \frac{PM}{P'M'}.$$

The triangles POM, $P'OM'$ are therefore similar, since the angles at P and P', contained between proportional sides, are equal. Hence the angles of the two triangles at O are equal, and the directions OM and OM' coincide. The path of the point M is therefore a straight line.

The motion is, moreover, uniform. From the same similar triangles,

$$\frac{OM}{OM'} = \frac{OP}{OP'} = \frac{t}{t'}.$$

The spaces described along the diagonal are proportional to the times, or the motion is uniform.

We see, therefore, that if the two motions which are applied to a particle at the same time are represented by the adjacent sides of a parallelogram, then the resultant motion will be represented by the diagonal of the parallelogram drawn through the intersection of these two sides.

Accelerations and velocities may be compounded in the same manner as uniform motions.

19. The Triangle of Motions. — Let two motions MA and MB be given to the particle at M at the same instant (Fig. 6). Construct upon these lines as adjacent sides the parallelogram $AMBC$ and draw through M the diagonal MC. MC represents the resultant of MA and MB both in direction and magnitude.

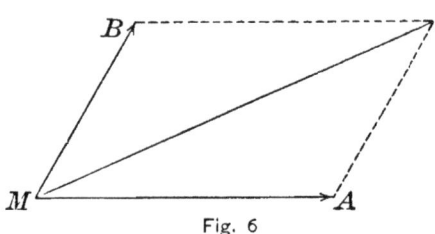

Fig. 6

Since AC is equal and parallel to MB, the three sides of the triangle MAC may represent the two component motions and their resultant. That is, if two sides of a triangle taken in order represent in magnitude and dircetion the two motions applied to a particle, their resultant will be represented by the third side of the triangle.

The line MC is called the *geometrical sum* of the two magnitudes MA and MB. Since this method may obviously be extended to any number of motions by means of a polygon of motions, we have the general law for the composition of motions, that *the magnitude of the resultant motion is the geometrical sum of the magnitudes of the component motions.*

20. Given Two Motions and the Angle between them, to find their Resultant. —

Let the two given motions, P and Q, be represented by the lines OA and OB respectively (Fig. 7); and let AOB be the given angle θ. Complete the parallelogram. It is required to find the value of the resultant R represented by the diagonal OC.

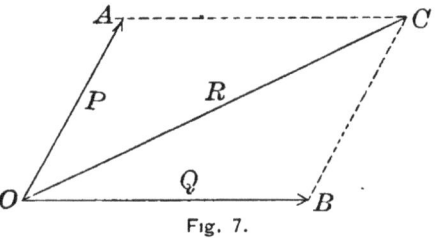

Fig. 7.

Since AC is equal to OB, it may equally well represent the motion Q. From plane trigonometry,

$$\overline{OC^2} = \overline{OA^2} + \overline{AC^2} - 2\,OA \cdot AC \cos A.$$

Substituting the values of the lines, and remembering that the angle A is the supplement of θ, we have

$$R^2 = P^2 + Q^2 + 2P \cdot Q \cos \theta \quad . \quad . \quad . \quad (13)$$

or the square of the resultant equals the sum of the squares of the two components plus twice their product into the cosine of the included angle.

Three particular values of θ give special results.

1. When $\theta = 0$. Then $\cos \theta = 1$, and

$$R^2 = P^2 + Q^2 + 2PQ, \text{ or } R = P + Q.$$

θ may be made zero by rotating OA around any point in it as O or A. If the rotation is around O the lines OA and OB finally coincide when $\theta = 0$. If OA is rotated about A, then when $\theta = 0$, OA is *parallel* to OB.

2.   When $\theta = 90°$. Then $\cos \theta = 0$, and

$$R^2 = P^2 + Q^2, \text{ or } R = \sqrt{P^2 + Q^2}.$$

3.   When $\theta = 180°$. Then $\cos \theta = -1$, $R^2 = P^2 + Q^2 - 2PQ$, and $R = P - Q$.

Again in this case θ may be made 180° by rotating OA counter clockwise around either O or A. In the first case the motions are oppositely directed along the same straight line, and in the second case they are oppositely directed and *parallel*. This result is of importance in compounding *parallel forces* under Kinetics.

Finally, when P equals Q with any angle θ

$$R = 2P \cos \frac{\theta}{2},$$

for the component of either motion in the direction of the resultant, which now lies midway between P and Q, is

$P \cos \dfrac{\theta}{2}$.

PROBLEMS.

1.   A body has impressed upon it two velocities of 25 and 20 metres a second at an angle of 45 degrees. Find the resultant velocity.

2.   A point has simultaneously impressed upon it three velocities of 70, 60, and 40 cms. a second. The angle between the first two is 60 degrees, and between the last two 45 degrees. Find the magnitude of the resultant.

21. Resolution in any Two Rectangular Directions. — The resolution of a motion, velocity, or acceleration in two rectangular directions is of more frequent occurrence than any other case. Let AD be a velocity v (Fig. 8) to be resolved into rectangular components. Let AB and AC be the two rectangular directions. Then the two components required are AB and AC.

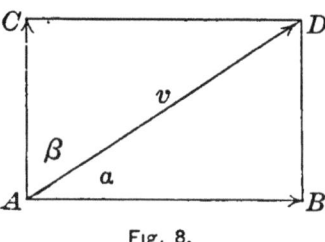

Fig. 8.

Let a and β be the direction angles which v makes with the two axes respectively. Then

$$AB = v \cos a \quad \ldots \quad \ldots \quad (14)$$
$$AC = v \cos \beta = v \sin a.$$

Hence to find the rectangular component of a velocity in any direction, multiply the given velocity by the cosine of the direction angle.

22. Extension of the Parallelogram of Velocities. — The composition and resolution of motions and velocities have thus far been limited to those varying as the first power of the time. The principle of the parallelogram may be extended to motions varying as any function of the time, provided both components are the same function. Thus both components may be a function of the square of the time, as in uniformly accelerated motion; or under due limitations they may vary as the sine or cosine of an angle which is proportional to the time, as in simple harmonic motion. The resultant will then be represented by the diagonal of the parallelogram, and will be the same function of the time as the components. But when one component is one function of the time and the other another function, then the extremity of the diagonal of the parallelogram, constructed on the two lines representing the

motions as adjacent sides, will be the position of the particle at the end of the time considered, but the particle
will not take the path of the diagonal to arrive at the
point.

The following topic will illustrate the principle.

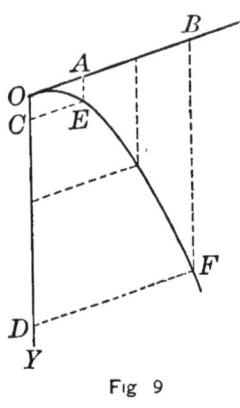

Fig 9

23. Path of a Projectile. — To re-
duce the problem to its simplest form
we shall assume that the projectile
meets with no resistance from the
atmosphere. Let v be the velocity of
projection in the direction OX (Fig.
9). This motion, which is uniform,
must be compounded with a uniformly
accelerated motion in a vertical direction OY.

Then in time t,

$$x = vt,$$
$$y = \tfrac{1}{2}gt^2.$$

Whence
$$x^2 = 2\frac{v^2}{g}y, \quad \cdots \cdots \quad (15)$$

an equation of the form $x^2 = 2py$, the equation of a
parabola.

Or we may reach the same conclusion geometrically as
follows:

Let OA and OB be distances along OX proportional to
times t and t'. Let OC and OD be the distances described with uniformly accelerated motion along OY in
the same time-intervals. Then

$$\frac{OA}{OB} = \frac{t}{t'}, \quad \text{and} \quad \frac{OC}{OD} = \frac{t^2}{t'^2}.$$

Whence
$$\frac{OC}{OD} = \frac{\overline{OA}^2}{\overline{OB}^2},$$

or the ordinates are proportional to the squares of the abscissas. This is a property characteristic of a parabola.

Equation (15) denotes a parabola referred to a tangent and a conjugate diameter. The path of the projectile is then a parabola tangent to the direction of the initial motion and having its axis vertical.

24. To find the Greatest Elevation and the Range (V., I, 201). — To discuss the motion of a projected particle it is more conven-
ient to refer it to rectangular axes. Let Ox (Fig. 10) be the horizontal axis passing through the point of projection O, and let Oy be the vertical through O. Let a be the angle between the direction

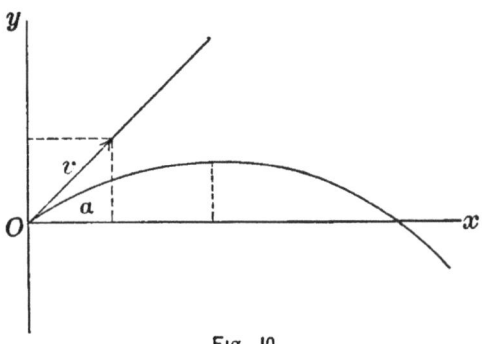

Fig. 10.

of projection and the horizontal axis. The velocity of projection v may be resolved into $v \cos a$ horizontal, and $v \sin a$ vertical. We have then to combine a uniform motion $v \cos a$ horizontally and a uniformly retarded motion along the vertical, with the initial velocity $v \sin a$ and the acceleration $-g$ due to gravity. The equations of the motion are

$$x = vt \cos a,$$
$$y = vt \sin a - \tfrac{1}{2}gt^2.$$

Eliminating t gives the equation of the path,

$$y = x \tan a - \frac{gx^2}{2v^2 \cos^2 a}.$$

This is the equation of a parabola of which the axis is vertical.

The summit of the curve is easily determined if we consider that at that point the vertical component of the motion is zero.

The vertical velocity at any time t is

$$v \sin a - gt.$$

Place this equal to zero and

$$t = \frac{v \sin a}{g}.$$

Substitute this value of t in the general expressions for x and y above and

$$x' = \frac{v^2 \sin a \cos a}{g} = \frac{v^2 \sin 2a}{2g}, \quad y' = \frac{v^2 \sin^2 a}{2g}.$$

The particle is then at the elevation $y' = \frac{v^2 \sin^2 a}{2g}$; it then descends along a branch symmetrical with the one traced during the ascent. The range a, or the horizontal distance to the point where it will again touch the horizontal plane through the point of projection, is therefore twice the value of x', or

$$a = \frac{v^2 \sin 2a}{g}.$$

Since v is a constant, the greatest range will correspond to an angle of elevation of 45°. The range is then $\frac{v^2}{g}$, or double the height which the projectile would attain if fired vertically with the same velocity v.

The range is the same for two angles of elevation equally distant from 45°, one above and the other below.

Practically the resistance of the air, which we have here neglected, will greatly modify these results, especially with large velocities of projection.

PROBLEMS.

1. A piece of ordnance under proof at a distance of 75 metres from a wall 7 metres high, burst, and a fragment of it, originally in contact with the ground, just grazed the wall and fell 3 metres beyond it on the opposite side. Find how high it rose in the air.

2. If a body be projected with a velocity of 33 metres a second from a height 22 metres above the ground at an angle of elevation of 30° with the horizontal, find when and where it will strike the ground.

25. Motion on an Inclined Plane. — This problem is often called Galileo's inclined plane, since he made use of it for the purpose of diminishing the effective component of the acceleration of gravity in determining the laws of falling bodies. Let the material particle be at a on the plane (Fig. 11), the angle of elevation of which is ϕ. The acceleration g to which the particle is subjected is in a vertical line, while the particle is constrained to move along the plane. Resolve the acceleration g into two components, one normal to the plane which is ineffective in producing motion, and the other parallel to the plane, the effective component. The latter is $g \sin \phi$.

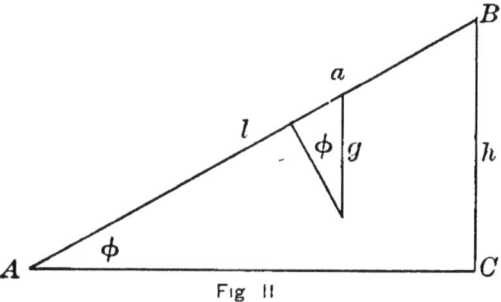

Fig. 11

Substitute this acceleration for a in equations (8), (9), (10), and

$$\left. \begin{array}{l} v = gt \sin \phi \\ s = \tfrac{1}{2}gt^2 \sin \phi \\ v^2 = 2gs \sin \phi \end{array} \right\} \quad \ldots \ldots (16)$$

From the figure $\sin \phi = \frac{h}{l}$. Hence when the body falls the entire length of the plane

$$v^2 = 2g\frac{h}{l}l = 2gh.$$

The velocity attained in descending the entire length of the plane is the same as in falling down the vertical height of the plane.

With the same conditions the second equation (16) gives

$$l = \tfrac{1}{2}g\frac{h}{l}t^2,$$

or

$$t^2 = \frac{2l^2}{gh}.$$

The time of descending a plane varies, therefore, as the length of the plane if the height remains constant.

PROBLEMS.

1. Find the gradient of a railway so that a carriage descending the plane by its own weight may travel 400 metres in the first minute; also find how far the carriage will go in the next minute, friction being neglected.

2. A body slides from rest down a sloping roof and then falls to the ground. The length of the slope is 6 metres, inclination to the horizon 30°, and the height of its lowest point from the ground $13\tfrac{1}{2}$ metres. Find the distance from the foot of the wall to the point where the body strikes the ground.

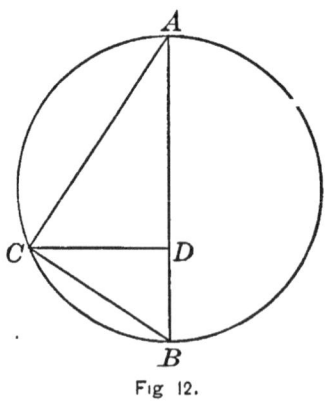

Fig 12.

26. **The Time of descent down any Chord of a Vertical Circle drawn through its highest point is Constant.** — Let AC (Fig. 12) be any chord drawn through the highest point of the vertical circle ABC. AB is a diameter. Draw CD perpendicular to AB.

Then the acceleration down AC is $g \sin ACD$. But by similar triangles

the angle ACD = the angle CBD.

Therefore $g \sin ACD = g \dfrac{AC}{AB}$.

But $s = \frac{1}{2}at^2$.

Substituting $AC = \frac{1}{2}g \dfrac{AC}{AB} t^2$.

Therefore $t^2 = \dfrac{2AB}{g}$, a constant.

The time of descent down any chord through A is therefore a constant and equal to the time of falling down the vertical diameter.

27. Uniform Circular Motion (T. and T., 9). — Hitherto, with the exception of the parabolic path of a projectile, the velocity has been assumed to vary in magnitude only; in other words, the acceleration has been confined to a single direction. But the velocity may vary also in direction. If the particle has a uniform rectilinear motion the acceleration is zero. If its velocity changes in magnitude without any change in direction, then the acceleration is positive or negative along the line of motion. If, however, the direction of the motion changes, then the particle has an acceleration one component of which is at right angles to its path. Thus if a particle move uniformly along AB (Fig. 13), while at

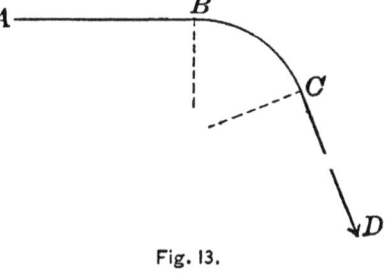

Fig. 13.

B it begins to describe a curved path from B to C, and from C on again moves uniformly along CD, then between B and C there is an acceleration normal to the path.

Acceleration should therefore be extended to mean rate of change of velocity in *any* direction in relation to the path.

In uniform circular motion the speed of the particle measured along the circumference is constant, while the acceleration is also constant, and is directed toward the centre. It is required to find the value of this centripetal acceleration.

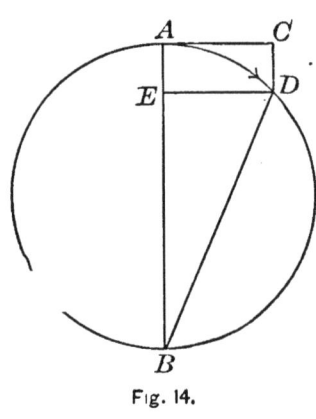

Fig. 14.

Let v be the velocity along the circumference. Then if AD (Fig. 14) be traversed in the interval t,

$$AD = vt.$$

The change of motion from AC, which is the direction of the path at A, to AD is AE. It will be noted, however, that the direction of the motion at D is not AD, but the tangent to the circle through D. The change in direction between A and D is therefore twice the angle CAD.

Since the centripetal acceleration is uniform,

$$AE = \tfrac{1}{2} ft^2,$$

where f is the required acceleration.

If now AD is an indefinitely short arc, then the triangles AED and ADB are similar, and

$$\frac{AE}{AD} = \frac{AD}{AB},$$

or
$$\overline{AD}^2 = AE \times AB.$$

Let r be the radius of the circle. Substituting the values of AD, AE, and AB, and

$$v^2 t^2 = \tfrac{1}{2} ft^2 \times 2r.$$

Whence
$$f = \frac{v^2}{r}. \quad \cdot \quad \cdot \quad \cdot \quad \cdot \quad \cdot \quad (17)$$

If T is the period of the complete revolution of the point, then

$$v = \frac{2\pi r}{T} \text{ and } v^2 = \frac{4\pi^2 r^2}{T^2}.$$

Substituting this value of v^2 in (17), and

$$f = \frac{4\pi^2 r}{T^2} \quad . \quad . \quad . \quad . \quad . \quad (18)$$

Since a radius connecting the point and the centre describes the angle 2π in the period T, the ratio $\frac{2\pi}{T} = \omega$ is called the *angular velocity*. Equation (18) may therefore be written

$$f = \omega^2 r. \quad . \quad . \quad . \quad . \quad (19)$$

PROBLEMS.

1. A mass of 1 gm. moves uniformly round a circle 40 cms. in diameter at the rate of 24 revolutions a minute. Compute the acceleration toward the centre.

2. Two equal masses, A and B, are connected by a string. The mass A describes a circle of radius 1 metre with uniform speed on the surface of a smooth horizontal table, while the other mass B is suspended against gravity by the string, which passes through a small, smooth hole at the centre of the table. Find the speed of A, assuming the acceleration of gravity to be 980 cms. per second.

28. Simple Harmonic Motion (T. and T., 19; D., 79; A. and B., 18; V., II, 260). — Simple harmonic motion is one of the most important motions which we have to consider, on account of its frequent use in sound, light, and alternating currents of electricity. While adhering, therefore, to an elementary method of treatment, we shall, nevertheless, discuss it somewhat in detail.

Simple harmonic motion is the apparent motion of a point, describing uniform circular motion, when viewed from a great distance in the plane of the circle. It is the component of the uniform circular motion at right angles to the line of sight; or it is the motion of the intersection of a diameter of the circle and a perpendicular from the moving point on this diameter. The simple harmonic motion is along a straight line with a maximum velocity at its middle point and zero at the extremities.

The satellites of Jupiter revolve in orbits which coincide very nearly with the plane of the ecliptic, or in a plane which passes nearly through the earth. Hence they appear to travel slowly backward and forward in nearly straight lines with simple harmonic motion.

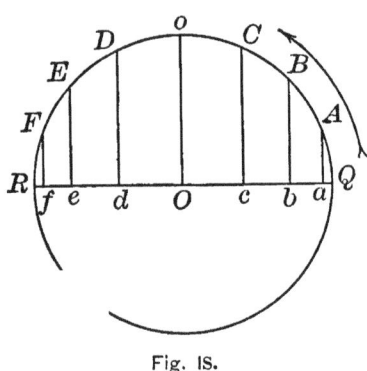

Fig. IS.

If the circle QAR (Fig. 15) represents the circle in which the point is moving, QR its apparent linear path, and A, B, C, D, etc., some of its successive positions, we may define its apparent motion by drawing perpendiculars from A, B, C, etc., and finding the points a, b, c, etc., to which the several positions of the point in the circle correspond. If the points on the circle are laid off at equal distances, then their corresponding projections on the diameter QR will not be equidistant; the distances Qa, ab, bc, etc., will, however, represent the spaces apparently traversed in equal intervals of time.

The motion along the diameter is not uniform, but is oscillatory, and the acceleration is directed toward the middle point of the diameter.

The circle is called the *auxiliary circle* or *circle of reference*, and its radius is the *amplitude* of the simple harmonic motion.

The *period* of the motion is the time of a complete revolution of the point around the circle of reference.

Motion from left to right is *positive*, and from right to left *negative*. Displacement to the right of the middle point is *positive*, and to the left *negative*.

The *phase* is the fraotion of a whole period which has elapsed since the particle last passed through the middle of its range in the positive direction.

When the particle is at Q it is said to be at its greatest positive *elongation*.

The *epoch* is the interval which must elapse from the point of reckoning time till the particle arrives at zero elongation going in the positive direction. In angular measure it is the angle described on the circle of reference during the time·interval defined as the epoch.

29. Acceleration in Simple Harmonic Motion proportional to Displacement. — Let the particle be at the point B in the circle of reference (Fig. 16). The *displacement* is the line OC. The acceleration toward the centre is $\dfrac{4\pi^2}{T^2} r$, since the partiele is moving uniformly around the circle. The problem is to resolve this centripetal acceleration into two rectangular components, parallel to OX and OY respectively.

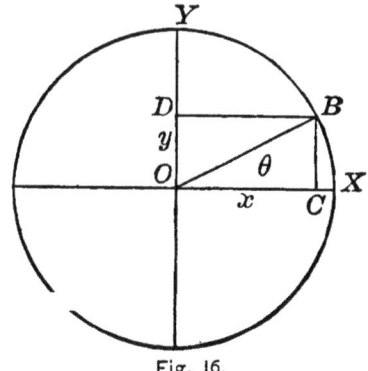

Fig. 16.

Denote the two component accelerations by f_x and f_y. Then, since the component in any direction is found by multiplying by the cosine of the direction angle,

$$f_x = - \frac{4\pi^2}{T^2} \, r \times \cos \theta,$$

$$f_y = - \frac{4\pi^2}{T^2} \, r \times \sin \theta.$$

Call the two displacements along the two axes x and y respectively; then

$$\cos \theta = \frac{x}{r}, \; \sin \theta = \frac{y}{r}.$$

Therefore
$$f_x = - \frac{4\pi^2}{T^2} \, r \times \frac{x}{r} = - \frac{4\pi^2}{T^2} \, x,$$

$$f_y = - \frac{4\pi^2}{T^2} \, r \times \frac{y}{r} = - \frac{4\pi^2}{T^2} \, y.$$

The first component varies as x and the second one as y; or the two components of the centripetal acceleration are proportional to the displacements of the points from the two rectangular diameters.

It is to be observed that when one of these components is a maximum the other is a minimum, or they differ in phase by a quarter of a period.

Now, since the *only acceleration* in uniform circular motion is directed toward the centre of the circle and is constant in value, it follows that uniform circular motion may be resolved into two simple harmonic motions at right angles to each other, of the same period and amplitude, and differing in phase by a quarter of a period.

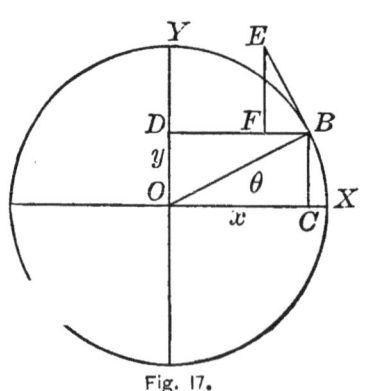

Fig. 17.

30. **The Velocity at any Point (A. and B., 19).** — To find the velocity of the simple harmonic motion at any point C (Fig. 17), resolve the uniform velocity V in the

auxiliary circle into two components parallel to the axes of X and Y.

Represent the velocity at B in the circle by BE. Complete the right triangle BEF. Then, since BE is a tangent, the angle E equals θ. Hence BF, the component parallel to the X axis, is

$$V_x = -\,V \sin \theta.$$

Also
$$V_y = V \cos \theta.$$

But
$$V = \frac{2\pi r}{T}:$$

Therefore
$$V_x = -\frac{2\pi r}{T} \sin \theta,$$

$$V_y = \frac{2\pi r}{T} \cos \theta.$$

When $\quad \theta = \dfrac{\pi}{2}, \sin \theta = 1,$ and $\cos \theta = 0.$

When one component of the velocity is a maximum the other is a minimum. The maximum velocity in simple harmonic motion is the same as the uniform velocity in the circle of reference.

It is important to observe that while the acceleration along the X axis is proportional to $\cos \theta$, the velocity is proportional to $\sin \theta$. When the point is at the middle of its range or course the acceleration is zero, but the velocity is a maximum; on the other hand, at the greatest elongation the acceleration is a maximum, and the velocity is zero. Starting at the extreme positive elongation at X, the acceleration decreases continuously toward the centre, while the velocity increases; but it increases at a continuously decreasing rate.

If the earth were spherical and of uniform density, and

if a hole could be drilled through its centre to the opposite side, then, neglecting resistance of the air and the rotation on the earth's axis, a heavy body dropped into the hole would descend with a diminishing acceleration and a constantly increasing velocity till it reached the centre; the acceleration would then become negative, and the velocity of the body would decrease, till at the opposite surface it would again come to rest, and would retrace its course, executing simple harmonic oscillations. Such is the motion of a point on a tuning fork or a pianoforte wire, and such is the motion of individual particles of air through which a simple fundamental tone is passing. The variations of electric pressure in the circuit of an alternating current dynamo follow approximately the same law of change.

PROBLEM.

A horizontal shelf moves vertically with simple harmonic motion, the complete period being one second. Find the greatest amplitude it can have in centimetres so that objects resting on it may remain in contact with it at its highest point, assuming g equal to 980.

31. To find the Resultant of Two Simple Harmonic Motions of the same Period and in One Line (T. and T., 21). — Let a and b be two points executing simple harmonic motions of the same period in the line OY (Fig. 18). Let

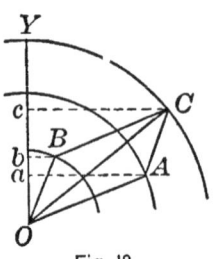
Fig. 18.

A and B be the corresponding uniformly moving points in the circles of reference.

On OA and OB describe a parallelogram and draw Aa, Bb, and Cc perpendicular to OY. Then the angle AOB is the difference in phase between the two harmonic motions to be compounded; and since they have the same period, this phase-difference has a fixed value, and the diagonal of the parallelogram remains of constant length, and makes constant angles with OA and OB. The point C therefore moves uniformly around a circle of radius OC, and the point c executes simple harmonic motion of the same period as a and b.

Moreover, bc equals Oa, since they are the projections on OY of the equal parallel lines BC and OA. Hence

$$Oc = Ob + bc = Ob + Oa.$$

But Oa, Ob, and Oc are the instantaneous values of the displacements of A, B, and C at any time t. The motion of c is therefore the resultant of the motions of a and b. It is a simple harmonic motion of the same period as that of the component motions. The resulting amplitude is the diagonal of the parallelogram constructed on the two component amplitudes, and making an angle with each other equal to their fixed difference of phase.

Let a parallelogram of cardboard, $OACB$, be cut out and be pivoted at O so as to turn freely counter clockwise. As the cardboard moves uniformly around, the magnitudes of the projections of the two sides OA and OB and the diagonal OC fluctuate according to a simple periodic law.

The resultant of the simple periodic motions, of which OA and OB are the amplitudes, and which have a fixed phase-difference equal to the angle AOB, is the simple harmonic motion of which OC is the amplitude.

The proposition is independent of the angle representing the fixed difference of phase, and therefore holds good when this difference is zero. In this last case not only is the resulting displacement equal to the algebraic sum of the component displacements, but the resulting amplitude equals the algebraic sum of the amplitudes of the component motions.

If the two amplitudes of the component motions are equal to each other, then the resulting amplitude equals twice the amplitude of either multiplied by the cosine of *half* the difference of phase (20).

When this difference of phase is half a period or π, then the resultant is zero.

When the two periods are nearly but not quite equal, while the amplitudes are still supposed equal, then the resulting amplitude passes slowly from twice that of either to zero and back again, in a time equal to that required for the faster to gain a complete oscillation on the slower.

This conclusion has important applications in sound and physical optics.

32. Composition of Two Simultaneous Circular Motions (D., 99). — Let the two uniform circular motions be in opposite directions around circles of equal radii (Fig. 19), and let them be of the same period.

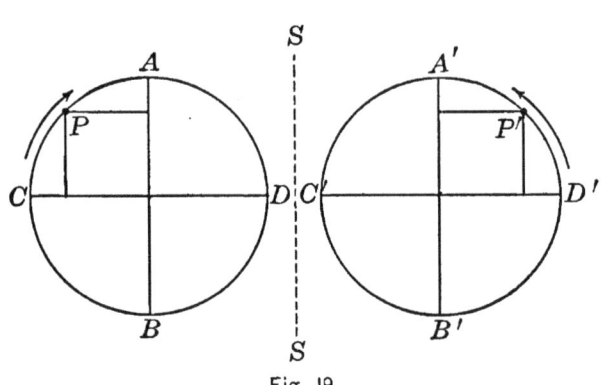

Fig. 19.

Resolve the two circular motions into simple harmonic components parallel to *AB* and *CD*. If now the two circular motions are conceived to be applied to the same particle, then since the harmonic components of the two circular motions parallel to *CD* are equal and oppositely directed, their resultant is zero. But those parallel to *AB* are in the same direction; and since they are simple harmonic motions and of the same period and phase, their resultant is a simple harmonic motion of the same period as the circular motions, and the amplitude is twice the common radius of the circles.

The resultant harmonic motion is in the plane of symmetry with respect to the two circular motions. The components *CD* differ in phase by half a period when referred to the plane of symmetry, while the components *AB* are in the same phase with respect to the plane of symmetry. When the two circular motions have the same period, a plane of symmetry can always be found, and the resultant simple harmonic motion will be parallel to this plane (Fig. 20).

If the periods of the two circnlar motions are not precisely equal, then one of the circular motions completes a revolution before the other, and the

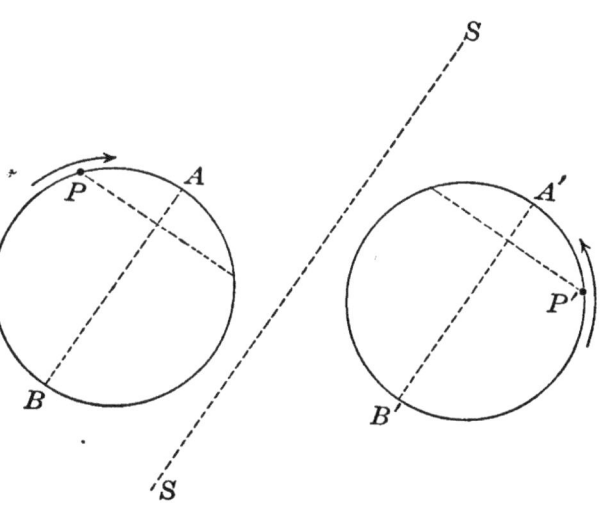

Fig. 20.

plane of symmetry revolves in the direction of the circular motion of shorter period. It makes one revolution while one circular motion gains a complete revolution on the other. This principle has an important application in the rotation of the plane of polarization of plane polarized light by such substances as quartz, solutions of sugar, and by the action of a magnetic field.

The converse of the above proposition is obviously true,

viz., that any simple harmonic motion may be resolved into two uniform circular motions in opposite directions, of the same period as the simple harmonic motion, and with radii equal to half its amplitude.

33. General Equations of Simple Harmonic Motion.—

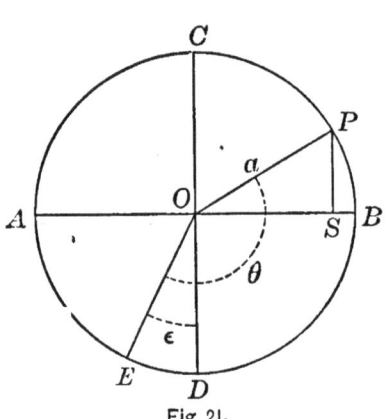

Fig 21.

Let x be the displacement parallel to AB (Fig. 21). It is required to find the value of this displacement. Let the point in the auxiliary circle be at P.

Then x equals OS, and

$$x = a \cos BOP,$$

or $\qquad x = a \cos \phi.$

But the definition of *phase* usually adopted renders it necessary to express x as a sine function of an angle.

Then $\qquad x = a \sin \left(\phi + \dfrac{\pi}{2} \right).$

Put θ for $\phi + \dfrac{\pi}{2}$, or the angle $D.OP$; then

$$x = a \sin \theta.$$

Let t be the time of describing the angle θ and T the period of a complete revolution, or of a complete simple harmonic oscillation. Then since the circular motion is uniform,

$$\frac{\theta}{2\pi} = \frac{t}{T},$$

and $\qquad \theta = \dfrac{2\pi}{T} t = \omega t.$

We may therefore write

$$x = a \sin \frac{2\pi}{T} t,$$

or $\qquad\qquad x = a \sin \omega t.$

But it is often necessary to reckon time or the angle θ from some other point than D, which corresponds to a displacement of zero. This is for the purpose of expressing difference of phase when two or more simple harmonic motions are compounded. Let time be reckoned from the fixed radius OE; the angle EOD is called the *epoch* ϵ. The angle θ then becomes EOP. Hence

$$x = a \sin (\theta - \epsilon),$$

or $\qquad\qquad x = a \sin \left(\frac{2\pi}{T} t - \epsilon \right).$

Since the number of complete oscillations per second n is the reciprocal of the period T, or $n = \frac{1}{T}$, we have

$$x = a \sin (2\pi n t - \epsilon).$$

Similarly we may write

$$V_x = \frac{2\pi}{T} a \cos \left(\frac{2\pi}{T} t - \epsilon \right),$$

and $\qquad\qquad f_x = - \frac{4\pi^2}{T^2} a \sin \left(\frac{2\pi}{T} t - \epsilon \right).$

Simple harmonic motion is therefore an oscillation in which the acceleration is proportional to the sine of an angle varying directly as the time. Since the sine of an angle has regularly recurring values, simple harmonic motion is a simple periodic function of the time; that is, the displacement, acceleration, and velocity are simple periodic functions.

CHAPTER III.

KINETICS.

34. Definition of Kinetics (T. and T., 52; B., 66). — Hitherto motion has been considered in the abstract, independently of the idea of matter and of the forces acting on it. But in Kinetics the forces producing the motion and the quantity of matter moved must be taken into account. For the velocity of a body depends not only on the magnitude of the forces acting, but also upon the quantity of matter set in motion. Kinetics treats of the action of forces in producing the motion of definite quantities of matter.

35. Mass, Volume, Density. — *Mass* is the quantity of matter contained in a body. It is expressed numerically in terms of the unit of mass, the gramme.

Volume is the space which a body occupies, expressed in cubic centimetres as the unit of volume.

Density is the mass of matter contained in unit volume. In the C.G.S. system it is the number of grammes of matter in a cubic centimetre. The mean density of a body may therefore be found by dividing its mass in grammes by its volume in cubic centimetres. The relation between density, mass, and volume may be written

$$d = \frac{m}{V} \text{ and } m = dV.$$

It is often convenient to employ the *volume containing unit mass.* Call it *s*. Then *d* and *s* are reciprocals of each other, or $d = \dfrac{1}{s}$.

36. Momentum. — The quantity of motion of a rigid body moving without rotation is considered to be made up of its mass and its velocity conjointly. It is called *momentum.*

$$Momentum = mv.$$

The whole motion of a body is the sum of the motions of its several parts. If the mass be doubled without changing the velocity, or if the velocity be doubled without changing the mass, then in both cases the quantity of motion is doubled.

Since in uniform motion $v = \dfrac{s}{t}$, we have $mv = \dfrac{ms}{t}$.

Speed is the rate of linear displacement; momentum is the rate of mass displacement.

37. Force. — Our conception of force is doubtless derived from muscular action. But great caution should be observed in transferring to the objective physical world any concepts derived from our sensations.

Force is said to act on a body when any change occurs in its state of rest or motion. Force is known by the change it produces in the motion of a body.

The *intensity of a force* is measured by the *acceleration* which it imparts to *unit mass.*

The *total magnitude* of a force is the product of mass and acceleration, or $F = m \dfrac{v - v_0}{t} = ma$. This acceleration may be a change in either the magnitude or the direction of the velocity. Weight is the total force of gravity, or $W = mg$.

The expression $m \dfrac{v - v_0}{t} = \dfrac{mv - mv_0}{t}$, from which it is seen that total force equals the time-rate of change of momentum. This is sometimes called the *acceleration of momentum.*

The rate of change of momentum of a falling body is constant, and in the vertical direction. But the rate of change of the momentum of a mass M, describing a circle of radius r, with uniform velocity v, is $M \dfrac{v^2}{r}$, and is directed toward the centre of the circle; that is to say, it depends upon a change of direction of the motion, not a change of speed (T. and T., 53).

An *impulsive* force acts for a very short period only, while a continued force acts for a longer or sensible period. The only difference is one of time. The time element is essential to produce any effect. When a hammer strikes an anvil it remains in contact with it for a measurable interval of time. So also when a bat strikes a ball the two remain in contact long enough for the ball to be brought to rest and to acquire motion in the opposite direction. Any force, however great, can produce only zero effect in zero time. Hence in estimating the action of a force the time element is of equal importance with the magnitude of the force. The word "impulse" takes both into account.

Impulse is the product of the force and the time during which it acts.

The *practical unit* of force is the *dyne*. It is that force which acting on a mass of one gramme for one second gives to it a velocity of one centimetre per second. It produces unit acceleration of unit mass.

The intensity of a force is measured by the acceleration

it produces. Hence we may substitute force for acceleration in the preceding propositions.

PROBLEMS.

1. What is the acceleration when a force of 36 dynes acts on a mass of 4 gms.? How far will the mass move in 10 seconds?

2. A force of 60 dynes acts on a body for one minute and imparts to it a velocity of 900 cms. a second. What is the mass of the body?

3. A mass of 500 gms. is whirled around at the end of a string 20 cms. long 3 times a second. What is the tension of the string, neglecting gravity?

4. Forces of 20, 30, 40 dynes act on a mass of 60 gms.; the angle between the first two is 45°, and that between the last two 60°. Find the space passed over in 10 seconds.

5. An engine winds a cage weighing 3,000 kilos. up a shaft at a uniform speed of 10 metres a second; what is the tension in the rope? What, if the cage move with a uniform *acceleration* of 10 metres a second?

6. An elevator starts to descend with an *acceleration* of 3 metres a second. Find the pressure on its floor of a man weighing 75 kilos. What would be his weight with respect to the elevator if it started to ascend with the same acceleration?

38. Newton's Laws of Motion (T. and T., 64 and 65). — The relations of motions and changes of motion to the forces producing them are expressed in Newton's three laws. These are to be regarded as physical axioms, which are not susceptible of rigorous demonstration. They are axiomatic to those only who have sufficient knowledge of physical phenomena to enable them to interpret their relations. The laws of motion must be considered as resting on convictions drawn from observation and experiment.

Law I. — Every body continues in its state of rest or of uniform motion in a straight line, except in so far as it may be compelled by impressed force to change that state.

Law II. — Change of motion is proportional to the impressed force, and takes place in the direction of the straight line in which the force acts.

Law III. — To every action there is always an equal and contrary reaction; or the mutual actions of two bodies are always equal and oppositely directed.

Newton defined the total motion of a body by the term *momentum.* By " *change of motion* " we should, therefore, invariably understand *change of momentum.*

Further, since the effect produced by a force depends upon the time during which it acts as well as upon the magnitude of the force, " *impressed force* " should always be interpreted as meaning *impulse.*

39. Discussion of the First Law of Motion (T. and T., 65-69 ; D., 5 ; B., 70). — The first law asserts that uniform motion in a straight line is as much the natural condition of a body as rest. The ideas embodied in the law are completely at variance with those of the ancient philosophers who asserted, on purely metaphysical grounds apparently, that circular motion is the only perfect one.

The term *rest* must be taken to mean rest with respect to other contiguous or related bodies. Absolute rest, no less than absolute motion, is unknown in nature. While, therefore, a body is at rest with respect to one body or to a system of bodies, it is at the same time in motion when another body or system of bodies is considered.

When no force acts on a body it persists in its state of rest *or motion* in relation to other bodies. Thus when a ball is fired from a cannon it continues to move, not because the force of the explosion follows it and keeps it moving, but because it meets with nothing capable of stopping it. It keeps moving *unless something stops it.* The question of the energy that it conveys from the source of motion is reserved for discussion in a later section.

Matter has no power in itself to change its condition of rest or of motion. Further, it offers resistance to any such change in proportion to the *mass* which it contains. This two-phased property of matter is expressed by the term *inertia*, and the first law of motion is often described as the *law of inertia.*

40. Inertia illustrated. — When a stream of water in a pipe is suddenly stopped by the closing of a valve, the shock which follows is due to the inertia of the mass of water which tends to keep it moving. This effect is utilized in the hydraulic ram. The shock due to the auto-matic closing of one valve opens another one leading into an air chamber. This latter valve opens under greater pressure than that due to the head of water in the pipe leading from the source. Hence a part of the water may be lifted to a higher level than the source of supply.

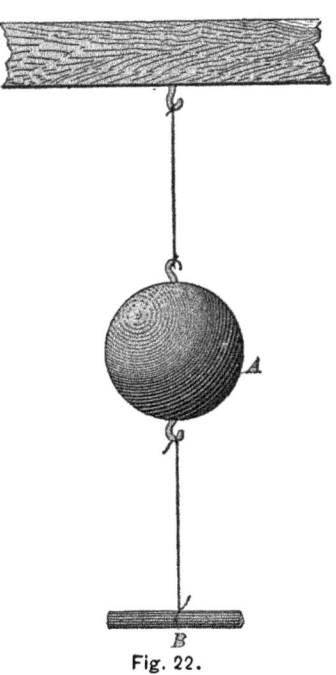

Fig. 22.

In earthquake shocks accom-panied by gyratory movements of the ground, heavy chimneys and isolated columns are sometimes left twisted around on their foun-dations. The inertia of the mass of the chimney or column causes it to remain fixed while the earth suddenly turns under it. A slower reverse gyration of the ground turns the column around with it.

Suspend by a string a heavy weight *A* (Fig. 22), and attach below by a piece of the same string the bar *B*. A *steady* pull downward on B will cause the string to break above A, because the tension of the upper string is equal to the weight of A in addition to the pull on B.

If, however, a *sudden* pull be applied to B the string will invariably break below A. The inertia of the weight A is so great that the lower string breaks before the stress reaches the upper one.

41. Discussion of the Second Law (L., 53). — The first law defines the condition under which a change of momentum takes place. The second law shows us first how a force may be measured. "Change of motion is proportional to the impressed force." Maxwell has re-stated this law in the following language: "*The change of momentum of a body is numerically equal to the impulse which produces it, and is in the same direction.*" The substitution of the word "equal" for "proportional" depends upon the definition of the unit of measurement. Impulse may therefore be placed equal to the change of momentum produced by it, or

$$Ft = mv - mv_0.$$

Whence
$$F = \frac{mv - mv_0}{t}.$$

This is the same expression for force as that found in Art. 37. When there is no change of momentum there is, therefore, no force. This is only the substance of the first law in another form. The first law is, therefore, involved in the second.

The second law shows further that the direction of the change of momentum always coincides with the direction of the impulse. This means that a force always produces its full effect on the motion of a body whatever its previous condition of motion or direction of motion. It implies also that if two or more forces act together on a body each produces its own change of momentum independently of the others.

We may therefore employ the same principles to compound forces that we have already employed for compounding motions. Forces may be compounded into a resultant or resolved into components by means of the parallelogram or triangle of forces in the same manner as motions. The resultant of any number of concurring forces is to be found by the same geometrical process as the resultant of any number of simultaneous velocities (T. and T., 67).

42. Discussion of the Third Law (M. and M., 77). — The third law expresses the fact that all action of force is of a dual character. All action between bodies or the parts of a system of bodies is of the nature of a *stress.* A stress is always a two-sided phenomenon. Every force, in fact, is one of a pair of equal and opposite ones — one component, that is, of a stress, either like the stress exerted by a stretched elastic cord, which *pulls* the two things to which it is attached with equal force in opposite directions, and which is called a *tension ;* or like the stress of a pair of railway buffers, or of a piece of compressed india-rubber, which exerts an equal *push* each way, and is called a *pressure.*[1]

When an elastic cord supports a weight, the stress on the cord, called tension, is equal in both directions. A stone attracts the earth with the same force that the earth attracts the stone. The existence of a stress in a medium in the case of gravitation has not yet been demonstrated, but it is thought to exist. The action between a magnet and a piece of iron is a mutual action, and the medium for

[1] Lodge's *Mechanics*, p. 55.

the exertion of this magnetic attraction is found to be in a state of strain. The magnet cannot exert a pull on the iron any greater than the iron exerts on the magnet. Two men pull at the opposite ends of a rope. The stress in the rope is obviously in both directions. It is no less certainly so when one end of the rope is tied to a post while one man pulls at the other end.

The same principle applies in cases where motion ensues. The centripetal force, which causes a rotating body to describe a circle with uniform velocity, is applied to the *rotating body* through the intermediary of the connection, visible or invisible, of the body with the centre. Corresponding to this is the opposite phase of this stress or the reaction, called the *centrifugal* force. This is not a force acting on the rotating body, but the reaction which the rotating body exerts on the centre. The centripetal and centrifugal forces are the opposite phases of the stress between the rotating body and the centre of rotation.

When any mutual action takes place between two bodies the momenta generated in opposite directions are equal; but the velocities are not equal unless the masses are equal. With unequal masses the velocities are inversely as the masses. While, therefore, the momentum of the gun is equal to that of the bullet, its velocity is very much less.

Considered only with respect to one portion of a system of bodies a stress is called *action;* with respect to the remainder of the system it is called *reaction.* The third law states that these two phases of a stress are always equal and in opposite directions.

PROBLEMS.

1. A gun weighing 3,000 kilos. and placed upon a smooth plane discharges a 30 kilogramme ball at an elevation of 30°. Find the relative velocity of the gun's recoil.

2. An inelastic mass of 900 kilos., moving with a velocity of 30 metres a second, meets an equal and similar mass moving 10 metres a second in the opposite direction. Find the velocity of the total mass after impact.

43. Work defined (B., 88 ; M. and M., 101 ; D., 39 ; Stewart's Conservation of Energy). — " Work is the act of producing a change of configuration in a system in opposition to a force which resists that change." Thus when a weight is lifted from the earth a change in the configuration of the weight and the earth is produced in opposition to the force of gravity which resists the change.

Work is measured by the product of the force and the displacement produced in the direction of the force. The amount of work is expressed as the product of two numbers, which represent respectively a force and a space.

Thus $$W = Fs \; ;$$

and since force equals the product of mass and acceleration,

$$W = mas.$$

When the displacement produced is not in the line of action of the force, but makes an angle a with that direction, then

$$W = Fs \cos a.$$

This may be described as the product of the force and the component of the displacement in the direction in which the force acts ; or the product of the displacement and the component of the force in the direction of the displacement. In one case it is the product of the force and the effective displacement; in the other the product of the displacement and the effective component of the force.

There is no work done by any force unless there is actual motion produced. Gravity does no work upon a weight at rest; it does work upon a falling weight.

The unit for the measurement of work in the C.G.S. system is the *erg.* It is the work done by a dyne through a distance of one centimetre.

Gravity gives to a gramme a velocity of approximately 980 cms. per second. It is therefore equal to 980 dynes. Hence if one gramme be lifted vertically one cm., the work done against gravity is 980 ergs; or, one erg of work is done in lifting $\frac{1}{980}$ gramme one cm.

Since work is the product of force and distance, it follows that

$$F = \frac{W}{s},$$

or *force is the linear rate of doing work.* The multiples of the erg sometimes employed are

The megalerg, or 10^6 ergs.
The joule, or 10^7 ergs.

Power or *activity* is the *time-rate of doing work.*

The unit of power employed in the C.G.S. system is the *watt,* which equals 10^7 ergs per second.

A horse power is a unit in the gravitational system, and is a rate of doing work equal to 33,000 foot-pounds per minute, or 550 foot-pounds per second.

To convert this into watts it is necessary to multiply the 550 by the ratio between a foot and a cm., then by the ratio between a pound and a gramme, which gives gramme-centimetres; and, finally, by the value in ergs of one gramme-centimetre, or 980. Hence

$550 \times 30.4797 \times 453.59 \times 980 = 746 \times 10^7$ ergs per second, or 746 watts.

One horse power is therefore equivalent to 746 watts.

44. Graphical Representation of Work (B., 89). — Since work is the product of force and space, it is evident that work may be represented numerically by an area. Thus when the force is of constant value the work done may be represented by the area of a rectangle, one side of which is numerically equal to the force and the adjacent side to the distance, or the component of the distance in the direction of the force (Fig. 23).

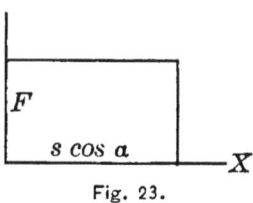

Fig. 23.

If the force increases from zero to a final value F, then the work done during this increase is the product of the mean value of the force and the displacement. It may then be represented by the area of a right triangle (Fig. 24),

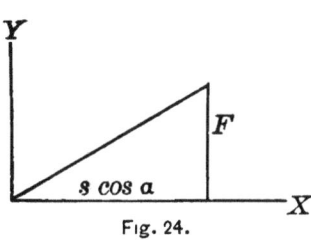

Fig. 24.

in which the base is the effective displacement, and the altitude the final value of the force F; for work then equals $\frac{1}{2}Fs \cos a$, which is the area of the triangle.

But in many cases, as in the cylinder of a steam-engine, the force or pressure varies according to a more complex law. If p is the pressure per unit area of the piston, and A is the area, then the total pressure is $P = pA$.

Let now x be a *small* distance through which the piston moves while the pressure p remains constant, then the work done by the gas during expansion is

$$w = pAx.$$

But Ax is the increase in the volume of the gas which we may denote by v.

Then $$w = pv,$$

or an element of the work done during a small movement of the piston is the product of the pressure per unit area and the small change in volume.

The total work done during a finite expansion of the gas may be represented by the area of the figure *ABba* (Fig. 25), in which the ordinates of the curve *AB* are the successive values of the pressure, and the abscissas represent the corresponding volumes of the gas. Take any small element of this area at *Aa.* Then the length of this strip is the instantaneous pressure *p*, and its width is the indefinitely small change of volume *v*. Hence its area is the element of the work done *w*, and the sum of all such elements is the total work done by the gas during the expansion. This is the area *ABba*.

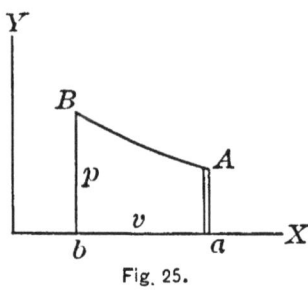

Fig. 25.

45. Energy (L., '79; T. and T., '73). — Whenever work is done on a body or a system of bodies, so as not merely to heat it, but in such a manner as to change the relative positions of its parts, then there has been conferred upon the system the capacity of doing work in its turn. If, for example, a mass of gas is compressed by a piston in a cylinder, work is done upon it as already explained. But the gas has now in turn acquired the ability to do work on the piston, both because of its higher temperature and the increased pressure to which it is subjected.

So when a steam-engine has lifted the weight of a pile-driver, it has done work on it against gravity. In its new position relative to the earth this weight has the capability of doing work; and when it is released, it descends, overcomes the resistance offered by the pile, and forces it into the ground.

When water is pumped up into an elevated reservoir work is done upon it; but the water thereby acquires the power of doing work, or of overcoming resistance, by means of a proper motor mechanism. It possesses something which it did not have at the lower level.

Work may be done upon a storage battery by means of a steam-engine and a dynamo-machine. The charged battery then has conferred upon it the power of doing work by the capacity which it has of furnishing a current of electricity to run an electric motor. It may churn the air by a fan, operate a printing press, run a street car, or propel an electric launch.

Consider some examples of a somewhat different character. Work is done upon a cannon ball by means of the pressure of the gases arising from the explosion of the powder. The ball acquires a high speed. It acquires more than that. It now possesses the capacity of overcoming resistance. By virtue of its mass and its motion it may demolish fortifications, pierce armor, or imbed itself in the nickel-iron plates of an ironclad.

When work is done by the steam on the piston of an engine, the heavy flywheel is made to revolve on its axis. Work is done upon it to give it motion. If the steam is shut off the engine will continue to revolve and may do work to the extent to which the massive flywheel now possesses the power of doing it.

In all these cases while work is being done upon the body *energy* is transferred from the active agent or the source to the system upon which work is done. The system or body which has acquired the power of doing work is said to possess *energy*. Strictly only the *available* energy is the power of doing work or of producing physical changes in other bodies; for a body may possess energy, in the form of heat for example, which cannot be made to do work, or is not *available*. Work may now be defined as the act of *transferring energy* from one body or system to another.

46. Potential Energy. — In the first three illustrations of the last section the body acquiring the energy exerts force or pressure. The compressed gas presses against the piston; the force of gravity pulls the weight of the pile-driver downward against the detent holding it; and the electric pressure of the battery is ready to produce a current as soon as the circuit is closed. During the transfer of energy from the working agent work is done against this force or pressure. The energy thus acquired is called *potential energy*. Potential energy is energy of *stress*. It is often called energy of position or static energy. In all cases of potential energy stress or force is present, but not motion; or at least the motion forms no part of the potential energy. The energy of an elevated mass, of bending, twisting, deformation, of chemical separation, and of the ether-stress in a magnetic field are all cases of potential energy.

47. Kinetic Energy. — In the last two illustrations of Art. 45 the effect of doing work upon the body is to give it motion. The energy which it thus acquires is called *kinetic energy* or energy of motion. Motion is the essential condition in a body possessing kinetic energy. Stress has no place in kinetic energy. But if a resistance is supplied work is done against it and the body loses kinetic energy.

When work is done both kinds of energy are usually present, and one of these forms of energy is passing into the other while the work continues. Thus the energy of steam in a boiler is partly potential, depending upon pressure, and partly kinetic, as heat; but the work that it does in the cylinder of an engine consists entirely in the immediate conversion of potential energy into kinetic energy, or energy of stress into energy of motion. Where-ever the motion is *with* the force speed increases, or potential energy becomes kinetic; when the motion is *against* the force kinetic energy becomes potential. Thus if a bullet be fired vertically upward, the motion is against the force of gravity, and the potential energy increases at the expense of the kinetic; when the bullet reaches its greatest elevation its energy is all potential. After that it descends, the motion is *with* the force, and the potential energy is again converted into the kinetic form.

The earth travels around the sun in an elliptical orbit with the sun at one focus. When the earth is nearest the sun, or at perihelion, it has its greatest orbital velocity and greatest energy of motion. As it recedes from the sun and passes toward aphelion, its acceleration toward the sun is not perpendicular to its path. This acceleration may be resolved into a normal and a tangential component. The first produces the curvature of the orbit, while the tangential component is a negative acceleration from perihelion to aphelion. The speed of the earth therefore diminishes, and its kinetic energy decreases, all the way from perihelion to aphelion. But its potential energy at the same time increases because the distance from the sun increases. At aphelion the kinetic energy is a minimum and the potential a maximum. The opposite transformation then takes place during the motion from aphelion to perihelion.

48. Energy is not Force. — It is a very common error to confuse energy and force. They are perfectly distinct and represent different physical concepts. When work is done on a body, so as to increase its available energy, the quantity of work done is equal to the gain in energy. Energy is therefore measured by work, and the unit of measurement is the erg, the same as the unit of work. The measure of energy always requires two factors, one of which only is a stress. Energy is always the product of one of several pairs of factors. These pairs do not always include force. Force is the space-rate of transferring energy, and is not the energy itself. Energy cannot be created, increased, or diminished by any natural or mechanical process. Force may be augmented to any extent by numerous mechanical devices. When a mass of matter moves without encountering opposition or resistance it carries with it a definite quantity of kinetic energy, but exerts no force. It produces neither motion nor change of motion till it transfers its energy to another body. Force is manifested only during this transfer.

49. Kinetic Energy in terms of Mass and Velocity. — In uniformly accelerated motion

$$v^2 = 2as.$$

Whence $$\tfrac{1}{2}mv^2 = mas.$$

But mass multiplied by acceleration is force. The second member of the last equation is therefore the product of force and distance, or work. It represents the work done upon a mass m to give to it the velocity v while working through the space s; and as the energy stored up is measured by the work done in its production, it follows that the energy of the mass m, moving with a velocity v, is $\tfrac{1}{2}mv^2$. Or conversely, a body m, moving with a velocity v, can do work against a resistance ma, through a space s, before coming to rest.

The same expression for the kinetic energy may be reached in another way. If a force F acts on a mass m during an interval of time t, then the measure of the effect is the impulse, or Ft.

By the second law of motion impulse is measured by the change of momentum. If the initial and final velocities of m are v_0 and v, then the change in momentum is $m(v - v_0)$. Therefore

$$Ft = m(v - v_0). \quad \ldots \quad \ldots \quad (a)$$

In uniformly accelerated motion the mean velocity is $\frac{1}{2}(v + v_0)$. It is also $\frac{s}{t}$. Hence

$$\frac{s}{t} = \frac{1}{2}(v + v_0). \quad \ldots \quad \ldots \quad (b)$$

Multiply (a) and (b) together, member by member, and we have

$$Fs = \tfrac{1}{2}m(v^2 - v_0^2) = \tfrac{1}{2}mv^2 - \tfrac{1}{2}mv_0^2.$$

But again Fs is the work done or the energy accumulated in the moving mass. The increase in the kinetic energy is therefore one-half the product of the mass and the difference of the squares of the initial and final velocities. If the initial velocity is zero, then

$$Fs = \tfrac{1}{2}mv^2$$

as before.

PROBLEMS.

1. A truck weighing 2,000 kilos. moves uniformly along a level surface with a velocity of 20 metres a second. Compute its kinetic energy.

2. Compare the kinetic energy of a ball having a mass of 15 gms and a velocity of 400 metres a second with that of the gun from which it was fired if the mass of the gun is 8 kilos.

50. The Conservation of Energy (M. and M., 103; Stewart's Conservation of Energy, 82). — In the admirable language of Maxwell, the principle of the conserva-

tion of energy is as follows: " The total energy of any material system is a quantity which can neither be increased nor diminished by any action between the parts of the system, though it may be transformed into any of the forms of which energy is susceptible." If some agent external to the system does work upon it, then its energy is increased by the amount of work done; if the system does work on any external resisting agent, then the system loses energy equal in amount to the work done by it. But if we include the given system and the external agent in one larger system, the energy of the total system is neither increased nor decreased by any action between them. By taking into account different parts of the physical universe successively, we finally reach the conclusion that the quantity of energy in the universe cannot be increased or diminished by any action of which we have knowledge. " The doctrine of the Conservation of Energy is the one generalized statement which is found to be consistent with fact, not in one physical science only, but in all. When once apprehended it furnishes to the physical inquirer a principle on which he may hang every known law relating to physical actions, and by which he may be put in the way to discover the relations of such actions in new branches of science."

The great doctrine furnished by chemistry is the indestructibility or conservation of matter; the great doctrine furnished by physics is the indestructibility or conservation of energy. These two entities, matter and energy, comprise all of the *physical* universe. Matter and energy.are indestructible by any power save that of the Almighty who created them. Is it preposterous or unscientific to put intelligence, the only other entity of which we have knowledge, in the same category?

51. Transformations of Energy. — While energy is indestructible it may assume almost innumerable forms, either potential or kinetic. A few examples of the conversion from energy potential to energy kinetic, or the reverse, have already been given. But all physical processes involve energy changes. A ceaseless series of such changes is therefore taking place in the orderly course of nature. Moreover, all machines are only instruments for the transformation of energy and the turning of it to account in effecting useful processes. A clock or a watch when wound up possesses a small store of potential energy which it gradually expends in doing the work of turning the mechanism against friction and the resistance of the air, and producing the sound of ticking. The striking mechanism of a tower clock receives its supply of energy for a week's service in the course of an hour. It distributes it over an entire week and sends it out hour by hour over an area of many square miles. The university clock requires 134,000 foot-pounds or 1816×10^9 ergs of energy to wind its striking side. It expends it by 484 strokes daily, including the chimes, or a total of 3388 strokes during the week. A watch in the same time divides up the energy given to it by daily windings into over 3,000,000 ticks.

An example of a different character is the energy in the form of chemical separation which is put into a rifle cartridge. The pull of the trigger operates to convert it into the kinetic energy of the bullet. The momentum is equally divided between the bullet and the gun, but not so with the energy; for kinetic energy varies as the square of the velocity. The kinetic energy therefore goes chiefly with the bullet.

An illustration of transcendent interest is found in the radiant energy received from the sun. The source of nearly all the energy which the earth possesses is the centre of the solar system. It comes to us through the medium of the ether as kinetic energy. Its absorption converts it into heat and warms the earth. In the marvelous chemistry going on in every leaf and plant, through the agency of the chlorophyll, it decomposes the carbon dioxide of the air and water, and is stored up in the cellulose and woody fibre as potential energy. In this form it may serve as food for man and animal, or may remain for ages securely locked up in the earth as coal and oil to furnish power and fuel for future generations.

In combustion or decay it again assumes the kinetic form and is dissipated beyond recovery as heat.

It warms the tropical ocean, lifts it in invisible watery vapor, transports it to distant continents, and pours it down as rain on mountains and elevated plains. Thence it descends as streams, and becomes the energy of water supplies.

The potential energy of the chemically separated fuel and oxygen may become mechanical power by means of the boiler and the engine. The potential energy of the elevated body of water may become mechanical energy by descending and driving the turbine wheel. By means of a dynamo-electric machine this mechanical energy may become the kinetic energy of an electric current. The conductor, or the ether surrounding the conductor, conveys it to distant points where electric lamps again convert it into light, or motors utilize it for any of the purposes to which power is applied. It is worthy of notice that electricity itself is not energy, but is only the medium by which the energy, which in every case has been derived from the sun, is distributed and utilized for man's comfort and convenience.

52. The Availability of Energy. — If the transformations of energy are attentively considered it will appear that the final form which it invariably assumes is diffused heat. Whenever the attempt is made to convert kinetic into potential energy, as by pumping water into an elevated reservoir, or by storing it up as energy of chemical separation in a storage battery, the conversion is always incomplete and partial.

It is impossible by any means at our command to effect a complete conversion of kinetic energy into the potential form. Some of it always escapes as heat, either through friction, radiation, conduction, the heating of electric conductors, or by some of the means by which energy escapes during conversion and transmission. But energy in the form of diffused heat is not available for further use. A heat-engine requires a hot body and a cold body in order that heat-energy may be utilized while passing from the hotter body to the cooler one.

In the same way all the processes of nature exhibit energy-changes on the way from the more available to the less available state.

Potential energy is the highly available form. It always tends to revert to the kinetic type, but in such a way that only a portion of the kinetic energy is available to effect useful changes either in nature or art. Hence, the energy of the solar system is becoming all the time less and less available.

No grander survey of the material universe, especially when considered from the point of view of energy-changes, has ever been made than the one described as the nebular hypothesis. Each celestial system is considered by itself. The nebular hypothesis traces it back to a mass of widely disseminated gaseous matter. It may not even at that period be faintly luminous; it possesses only the inherent quality of gravitation, and therefore of potential energy. Slowly the widely extended mass gravitates toward its centre of mass. The parts of the system fall toward one another, and its potential energy thereby begins to suffer conversion into the kinetic energy of heat and light. As long as the falling together and the contraction of the mass continue, so long is energy of position transformed into energy of motion. It is easily demonstrable that the temperature of the contracting mass continues to rise as long as it remains gaseous. The radiation of energy into space then goes on. The contraction of the mass means, therefore, the dispersion of the energy. The potential energy of the diffused nebulous matter is convertible and available; the converted kinetic energy radiated into boundless space is chiefly unavailable.

Of this energy radiated from the sun the earth receives about one two-thousand-two-hundred-millionth part $(\frac{1}{2,200,000,000})$. And of this small fraction the portion stored up in vegetation and coal is only an infinitesimal part. Nature is therefore prodigal of energy, not economical; and the small store that the earth does retain ultimately becomes diffused and radiated heat.

It is thus apparent that the energy of the solar system is running to waste or becoming unavailable.

Unavailable energy is called *Entropy*. The inevitable conclusion is that entropy tends toward a maximum.

All the operations of nature transform energy from the available into the unavailable form. Following backward therefore the history of natural operations brings us to a far-distant period when all the energy of the solar system, and indeed of the physical universe, was in the available form. This must have been before any natural operation took place. "In the beginning," then, points to the period when all energy was available.

With no less certainty physical science points to a time when entropy shall be a maximum. All the processes of nature must then cease. Even the earth itself, as lifeless as the moon, can no longer circle round the glowing sun, but both and all together, in one dead mass, must hang in everlasting silence in the boundless night of space. The marvellous mechanism will then have run down, and no further motion or life-process will be possible unless some new order intervenes of which we have no knowledge or conception.

PROBLEMS.

1. A mass of 20 kilos. moving with a velocity of 16 metres per second, overtakes a second mass of 32 kilos. moving with a velocity of 12 metres per second; find the common velocity after impact. Calculate the loss of kinetic energy, both bodies being inelastic.

2. To a vertical axle revolving 100 times per minute there is attached at its lower end a mass of 75 kilos. by means of a cord 1.5 metres long. What angle does the cord make with the vertical, and what is the tension in the cord?

3. A tooth in the blade of a reaper describes a simple harmonic motion of 4 cms. amplitude in a period of one-seventh of a second. What is its maximum velocity and its maximum acceleration per second?

4. Find the velocity with which a body should be projected down an inclined plane, so that the time of descending the plane shall be the same as the time of falling through the vertical height of the plane.

CHAPTER IV.

KINETICS (*Continued*).

53. The Moment of a Force. — "The moment of any physical agency is the numerical measure of its importance." The moment of a force with respect to a point is the measure of the tendency of the force to produce rotation about the point. It is measured by the product of the line representing the force and the length of the perpendicular drawn from the point to the direction of the force. Let *O* (Fig. 26) be the point about which the body acted upon by the force *AB* is constrained to rotate. The moment is then

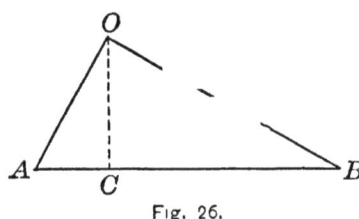

Fig. 26.

$$F \times OC = AB \times OC.$$

This is double the area of the triangle *ABO*. The line *OC* is called the *acting distance* or *lever arm*, and the point *O* is the *origin* of moments. The moment of a force producing rotation in one direction is positive, and the moment producing rotation in the other direction is negative. In the figure the rotation is supposed to be counterclockwise. This is often called the positive direction. A rotation clockwise would therefore be negative. It is not necessary always to observe this conventional rule.

PROBLEMS.

1. *ABCD* is a square of 5 metres on each side. Forces of 50, 60, and 70 dynes act along *AB, AC*, and *AD*. Find the moment of each force about *D*.

2. Find the moment of each of the forces in problem 1 about the middle point of *AC*.

54. The Moment of the Resultant equals the Algebraic Sum of the Moments of the Components. — Let the origin *O* fall outside the angle included between the lines representing the two forces *P* and *Q*. The moments are all then of the same sign. Complete the figure as shown in the diagram (Fig. 27). Then *p, q,* and *r* are the acting

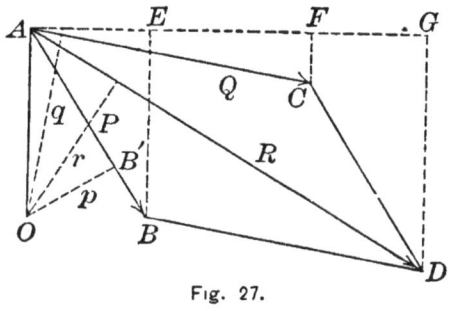

Fig. 27.

distances of the two forces and their resultant.

From the similar triangles *OAB′* and *ABE*,

$$\frac{AE}{p} = \frac{AB}{AO}, \text{ or } AE = \frac{AB \times p}{AO}.$$

In the same way

$$\frac{AF}{q} = \frac{AC}{AO}, \text{ or } AF = \frac{AC \times q}{AO}.$$

Also $\dfrac{AG}{r} = \dfrac{AD}{AO}$, or $AG = \dfrac{AD \times r}{AO}$.

But *AE = FG*, since they are projections on the same line of equal and parallel lines. Therefore

$$AG = AF + AE.$$

Substitute the values of *AG, AF*, and *AE*, and multiply through by *AO*. Then

$$AD \times r = AB \times p + AC \times q,$$

or $$R \cdot r = P \cdot p + Q \cdot q.$$

The moment of the resultant equals the sum of the moments of the components.

If O lay within the angle BAC, one of the moments would be of a different sign from the others. In that case the moment of R would equal the arithmetical difference of the other two moments.

Since the proposition is true for two forces and their resultant, it is also true for this resultant and a third force, and so on indefinitely. It is, therefore, true for any number of concurring forces.

A case of particular importance occurs when the origin O is on the line of the resultant or this line produced.

The moment of the resultant is then zero ; and the sum of the positive moments is equal to the sum of the negative moments, or the tendency to rotate in one direction equals the tendency to rotate in the other direction. The equation of equilibrium is then written by placing the algebraic sum of the moments equal to zero.

If any line in a body is fixed the resultant of all the forces passes through this line if there is no rotation. Hence the sum of all the moments with respect to a point on this fixed line is zero.

55. Parallel Forces. — *First, two forces.* Let P and Q be two parallel forces in the same direction in Fig. 28, and in opposite directions in Fig. 29. Then by Art. 20 the resultant is $P + Q$ in the first case, and $P - Q$ in the second, P being greater than Q. Let C, on the line AB joining the points of application of the two parallel forces, be the point of application of the resultant. Take C as the origin of moments and draw DE perpendicular to the direction of P and Q.

Then since the origin is on the line of the resultant, the moment of the resultant is zero, or

$$P \times DC - Q \times EC = 0.$$

Whence

$$\frac{P}{Q} = \frac{EC}{DC}.$$

But

$$\frac{EC}{DC} = \frac{BC}{AC}.$$

Therefore

$$\frac{P}{Q} = \frac{BC}{AC},$$

or the lines joining the points of application of the two forces with that of the resultant are inversely proportional to the forces.

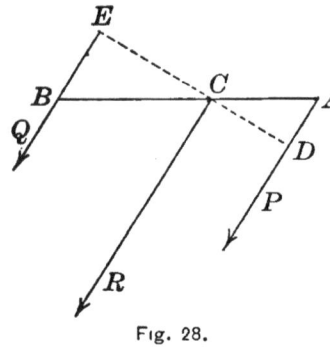

Fig. 28.

Also

$$\frac{P}{P+Q} = \frac{BC}{BC+AC},$$

or

$$\frac{P}{R} = \frac{BC}{AB} \text{ (Fig. 28)};$$

and

$$\frac{P}{P-Q} = \frac{BC}{BC-AC},$$

or

$$\frac{P}{R} = \frac{BC}{AB} \text{ (Fig. 29)}.$$

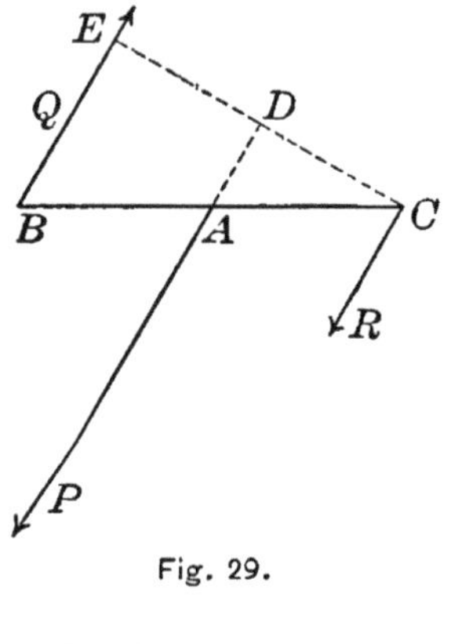

We may therefore write

$$P : Q : R :: BC : AC : AB.$$

We may then say that of two parallel forces and their resultant, each is proportional to that part of the line joining their points of application which is included between the other two.

Fig. 29.

Second, any number of parallel forces. Let there be any number of parallel forces, P, P', P'', etc. Then

$$R = P + P' + P'' + \text{etc.},$$

each force with its proper sign.

Let O be any point on a right line cutting the directions of the parallel forces at right angles, and let x, x', x'', etc., be the distances from the intersections of the forces with this line to the point O. Also let X be the distance of O from the intersection of the resultant R with the same line. Then from the principle of moments

$$RX = Px + P'x' + P''x'' + \text{etc.}$$

This may be written

$$RX = \Sigma Px,$$

where the symbol Σ means " the sum of such terms as," all the terms being of the same form.

Finally

$$X = \frac{\Sigma Px}{\Sigma P}.$$

This determines the line of action of the resultant.

PROBLEMS.

1. A stick of timber of uniform cross-section is carried by three men, one at one end and two by means of a bar placed crosswise under the stick. Where must the bar be placed that each man may carry one-third the weight?

2. A horizontal bar AB, 3 metres long, has one end B attached to the vertical side of a building. The other end A is supported by a rope tied to it and to the building at a point 4 metres above B. The bar supports a weight of 50 kilos. at its middle point. Find the tension in the rope (by moments).

3. Two men carry a weight of 50 kilos. slung on a light pole 280 cm. long. If the weight be placed at a distance of 100 cm. from one end, what weight does each man carry ?

56. Couples. — When the two parallel forces are equal and oppositely directed, their resultant is zero. Such a pair of forces constitute a *couple*. Their resultant is zero so far as motion of translation is concerned, in which all points of the body move in parallel straight lines, or any line drawn within the body remains fixed with respect to the body and moves parallel to itself.

In Fig. 29
$$\frac{P}{R} = \frac{BC}{AB}.$$

If now R is zero the first member of the equation is infinity; therefore BC is infinite in value, since AB is not zero. The resultant is zero, and its point of application is at an infinite distance.

Such a pair of forces may cause a body to revolve around an axis. The *moment of a couple* is the product of one of the forces and the perpendicular distance between their lines of action.

One couple is in equilibrium with another when the moment of the one equals the moment of the other, and their directions of rotation are opposite.

PROBLEMS.

1. Four forces are represented in direction, magnitude, and line of action by the sides of a square taken in order; prove that the sum of their moments about every point of the square is a constant.

2. Prove that the forces in the last problem are equivalent to a couple.

3. Prove that the moment of this couple is numerically equal to twice the area of the square.

57. The Lever. — A *lever* is any rigid rod or bar, the weight of which may be neglected, free to turn about a fixed point called the *fulcrum*. The problem is to find the

conditions of equilibrium for two forces tending to turn
the lever in opposite directions about the fulcrum.

For this purpose it is necessary to recall that a *resultant* is a single force which will produce the same effects
as the several forces compounded. Hence if a force be
applied to the body equal to the resultant of all the forces
acting and opposite to it in direction, then equilibrium must
ensue. For if we assume the several forces replaced by
their resultant and a force applied equal and opposite to
this resultant along the same straight line, the final resultant would be zero, or no motion would ensue. Each
force would then be equal and opposite in direction to
the resultant of all the others.

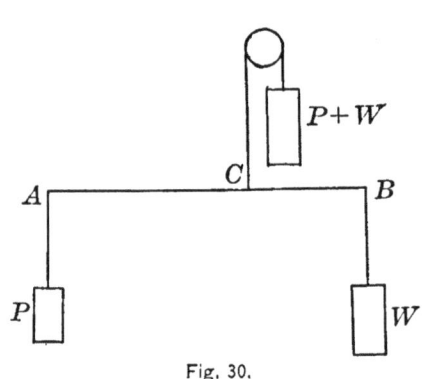

Fig. 30.

Let two weights, P and
W (Fig. 30), be applied to
the two ends of the lever,
and let the lever be supported at the point of application
of their resultant by a cord
passing over a pulley without friction, and carrying a
counter weight $P + W$
Then the several weights
will be in equilibrium. Either A, B, or C may be conceived to be fixed and the equilibrium will be maintained.

Thus if C be fixed it becomes the fulcrum, and by the
principle of moments (54)

$$\frac{P}{W} = \frac{BC}{AC},$$

the moment of the resultant with respect to C being zero.
This constitutes a lever of the first order.

If A be fixed it is the fulcrum, the resultant of W and
$P + W$ passes through A, its moment is zero,

and
$$\frac{W}{P + W} = \frac{AC}{AB},$$

or power and weight are inversely as their acting distances.

If W is the power, the lever belongs to the second order.

If $P + W$ is the power, the lever is of the third order.

The relation of power and weight, expressed above, applies to the several orders alike.

58. Principle of the Lever by the Theory of Work. — Suppose the system, represented by Fig. 31, to suffer a small displacement, so that A and B take the positions A' and B'. Then the positive work done on one side by W equals the negative work done on the other side against P.

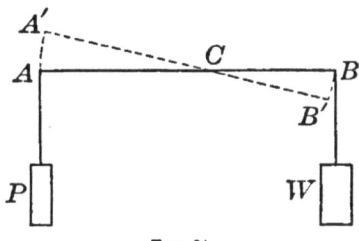

Fig. 31.

If the displacement is small, the arcs AA' and BB' are the distances over which the work is done on the two sides. Hence
$$P \cdot AA' = W \cdot BB',$$

or
$$\frac{P}{W} = \frac{BB'}{AA'}.$$

But similar arcs are to each other as their respective radii. Therefore
$$\frac{P}{W} = \frac{BC}{AC},$$

or power and weight are to each other inversely as their lever arms or acting distances.

59. The Inclined Plane. — Since accelerations are proportional to forces, the relation already found to exist between the accelera- tion of gravity in free fall and the acceleration down an inclined plane also ap- plies to power and weight, when the power is applied parallel to the face of the plane AC (Fig. 32).

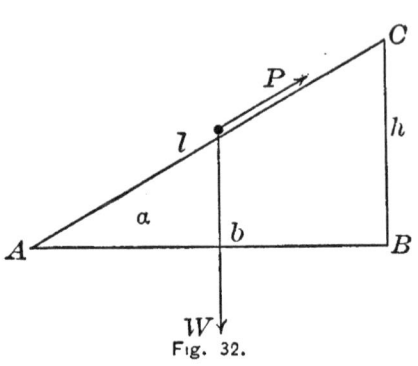

Fig. 32.

If g' is the acceleration parallel to the face of the plane,

$$g' = g \sin a,$$

or $$mg' = mg \sin a.$$

But mg' is the force or weight P and mg is the weight W. Hence

$$P = W \sin a.$$

Substituting $$P = W\frac{h}{l}, \text{ or } \frac{P}{W} = \frac{h}{l}.$$

The same result may be obtained by means of the prin- ciple of work, for the work done by P along the whole face of the plane AC equals the work done against gravity in lifting W through BC. Hence

$$P \cdot l = W \cdot h,$$

or $$\frac{P}{W} = \frac{h}{l}.$$

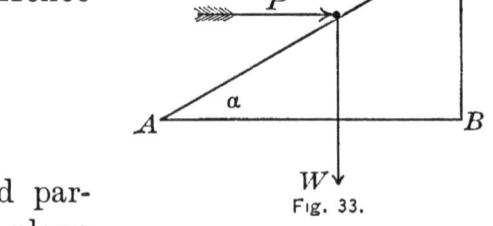

Fig. 33.

If the power is applied par- allel to the base of the plane (Fig. 33), then its component parallel to the face of the plane must equal the component of W in the same direction, or

$$P \cos a = W \sin a.$$

Hence
$$\frac{P}{W} = \frac{\sin a}{\cos a} = \tan a = \frac{h}{b}.$$

If the principle of work is applied, then the distance worked over must be multiplied by the component of the force in the direction of the displacement to find the work done. Again assuming that the weight is moved from the bottom to the top of the plane, we have

$$P \cos a \, l = Wh.$$

But
$$\cos a = \frac{b}{l}.$$

Hence
$$P \frac{b}{l} l = Wh,$$

or
$$\frac{P}{W} = \frac{h}{b}, \text{ as before.}$$

The inclined plane serves to increase the time of doing a given amount of work, or conversely of decreasing the rate of doing the work, that is, decreasing the *power* required. In the approaches to the St. Gotthard tunnel the necessary rise to reach a given elevation is distributed over a long spiral inclined plane, lying partly within the mountain as a tunnel and partly on its face, and of such length that the power required to lift the train is reduced to that which the engines can supply.

60. The Sensibility of the Balance (A. and B., 76). — The balance is an instrument for the comparison of equal masses of matter. It consists of a light, trussed beam, so as to have the required stiffness with the least weight, and it is supported at its middle point by means of a " knife-

edge " on agate planes.　At each end is suspended a scale
pan, and the two pans should be of equal weight.

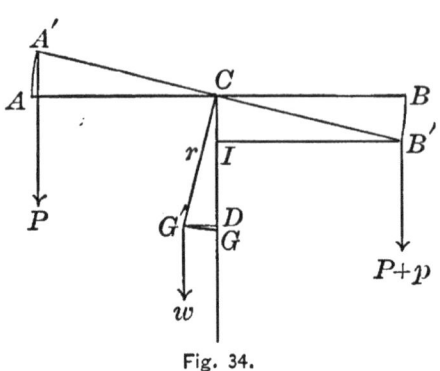

Fig. 34.

Let the three points _A_,
_B_, and _C_ (Fig. 34) be in
the same straight line.　_A_
and _B_ are the knife-edges
supporting the scale pans,
and _C_ is the knife-edge
on which rests the beam.
The centre of gravity of
the beam is at _G_, and the
weight of the beam is _w_.

Let a weight _P_ be placed
in one scale pan and _P_ + _p_ in the other.　These weights
include those of the pans.　The two arms of the balance
should be equal to each other.　Then, upon producing a
small displacement, so that the beam takes the position
_A'B'_, the two lever arms remain equal to each other, so
that _P_ and _P_ have equal moments and counterbalance
each other.　They may therefore be omitted from the
equation of equilibrium.　It is necessary then to take
into account only the moments of the small excess weight
_p_ and the weight of the beam _w_.　These produce rotation
in opposite directions around the fixed point _C_ as the
origin of moments.

The sensibility of the balance is proportional to the
angular displacement of the beam with a given small difference _p_ in the load.

When the beam assumes its position of equilibrium
_A'B'_, the moment of _p_ on one side equals the moment of _w_
on the other.　Therefore

$$p \times B'I = w \times G'D.$$

Let CG be called r. If l is the length of each arm of the balance, then $B'I$ equals $l \cos \theta$ and $G'D$ equals $r \sin \theta$. Substituting

$$p \times l \cos \theta = w \times r \sin \theta.$$

From which

$$\frac{\sin \theta}{\cos \theta} = \tan \theta = \frac{pl}{wr}.$$

The tangent of the angular displacement, or if the angle is small the sensibility, varies directly as the length of the beam, and inversely as the weight of the beam and the distance between its centre of gravity and the knife-edge or axis of suspension.

If the three points, A, B, C, are not in the same straight line, the sensibility changes with the load. When C is above AB, the sensibility is diminished; for when the beam is displaced the lever arm of the higher end of the balance beam becomes longer than that of the lower end, and the moments of the two weights P and P become unequal to each other. The difference in moments increases with P and tends to diminish θ.

When C is below AB the sensibility is increased. The lever arm of the lower end of the beam becomes greater than the other, and the difference in moments increases θ.

The deflection of the beam under load raises the point C with respect to the line AB. Hence increase of load may produce first an increase and then a decrease of sensibility.

61. Double Weighing. — The process of double weighing serves not only to determine the true weight when the arms of the balance are unequal in length, but also to determine the ratio of the arms.

Let l and l' be the lengths of the arms AC and BC.

Let the body of weight w be first suspended from A and counterbalanced by a weight w' in the other scale pan. Then let it be placed in B and counterbalanced by weight w''.

Then for the first operation

$$lw = l'w'. \quad \cdot \quad \cdot \quad \cdot \quad \cdot \quad \cdot \quad \cdot \quad \cdot \quad (a)$$

For the second

$$l'w = lw''. \quad \cdot \quad \cdot \quad \cdot \quad \cdot \quad \cdot \quad \cdot \quad \cdot \quad (b)$$

Multiplying together (a) and (b) member by member

$$ll'w^2 = ll'w'w''.$$

Therefore $\qquad w = \sqrt{w'\,w''}.$

When the two are very nearly of the same length, and w' differs but little from w'', the true weight may be found very approximately by taking the half sum of w' and w''.

To find the ratio of the two arms it is convenient to proceed somewhat differently. Take two weights of the same nominal value W and W' and obtain equilibrium with one in each pan. Suppose that when W is placed in the left-hand pan w' must be added to it, and when in the right-hand pan w'' must be added to secure a balance. Then

$$l\,(W + w') = l'\,W',$$
$$l\,W' = l'\,(W + w'').$$

Therefore $\qquad l^2\,W'\,(W + w') = l'^2\,W'\,(W + w''),$

or $\quad \dfrac{l}{l'} = \left(\dfrac{W + w''}{W + w'} \right)^{\frac{1}{2}} = \left(\dfrac{1 + \dfrac{w''}{W}}{1 + \dfrac{w'}{W}} \right)^{\frac{1}{2}} = 1 + \dfrac{w'' - w'}{2\,W}\text{ nearly.}[1]$

Both w' and w'' must be taken with the proper sign.

---

[1] Kohlrausch's *Physical Measurements.*

PROBLEM.

Left Arm.	Right Arm.
$W = 200$ gms.	$W' + .008073$
$W' + .0005$	$W = 200$

Find the ratio of l to l'.

62. Centre of Inertia (A. and B., 44). — If we consider any system of *equal* material particles, their *centre of inertia* is the point whose distance from any plane is the average distance of the several particles from that plane.

When the body or system consists of a finite number of parts which are not equal, but the masses of which are known, then the distance of the centre of inertia of the whole body from any plane may be found by taking the sum of the products of the several masses by their respective distances from the plane and dividing by the sum of the masses. The distance of the centre of inertia is thus the average distance of all the masses from the plane.

Conceive now the system of particles to be acted on by a system of parallel forces proportional to the masses of the particles. Then if the particles be supposed collected at the centre of inertia, or centre of mass, and to be acted on by the resultant of all the parallel forces, the same motion will be produced as if all the separate forces acted on the separate particles. The motion will be motion of translation without rotation.

When the parallel forces are due to gravity the point defined above is called the *centre of gravity*. The centre of inertia and the centre of gravity are identical when the forces of gravity are strictly parallel and proportional to the several masses constituting the body.

The distance X of the centre of inertia of any body or

system of bodies from a given plane may be expressed in the form of a general formula,

$$X = \frac{\Sigma mx}{\Sigma m}.$$

It may be demonstrated as follows : Let MM' (Fig. 35)

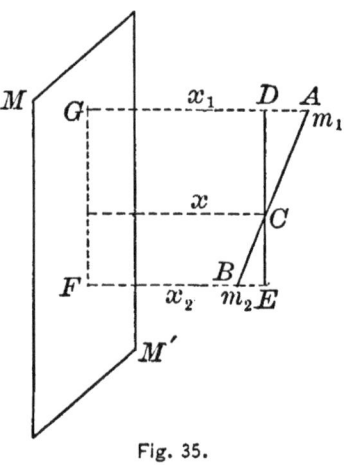

Fig. 35.

be the plane of reference. Let the particles m_1, m_2, be situated at A and B, distant respectively x_1 and x_2 from the plane. Let C be the centre of mass of the two particles; that is, the centre of two parallel forces at A and B proportional to m_1 and m_2. Through C draw DE parallel to FG; F and G are the points of intersection of the perpendiculars from m_2 and m_1 with the plane. Let x be the distance of C from the plane.

Then

$$\frac{m_1}{m_2} = \frac{BC}{AC} = \frac{BE}{AD},$$

or $$m_1 \times AD = m_2 \times BE.$$

But $$AD = x_1 - x, \text{ and } BE = x - x_2.$$

Hence $$m_1 (x_1 - x) = m_2 (x - x_2).$$

Transposing, $$x (m_1 + m_2) = m_1 x_1 + m_2 x_2.$$

The same process may be applied to the sum of m_1 and m_2 and a third particle m_3, and so on indefinitely. We may therefore write generally

$$X\Sigma m = \Sigma mx,$$

or $$X = \frac{\Sigma mx}{\Sigma m}.$$

If in the figure the plane of reference should pass through C, x would be zero. In general when the plane of reference passes through the centre of inertia, so that X is zero, we have

$$\Sigma mx = 0.$$

63. Centre of Inertia of a Triangle. — Let the triangle be of uniform thickness and density. Divide it into a *very large number* n of small areas by equidistant lines drawn parallel to the base. From the apex C (Fig. 36) draw CD to the middle point of the base. It passes through the middle points of all the narrow elements of the triangle, which are their centres of inertia. Their centres of inertia all lie on the line CD, and therefore the centre of inertia of the entire triangle must lie on the same line. Let it be at G. The problem is to find the distance of G from the apex C.

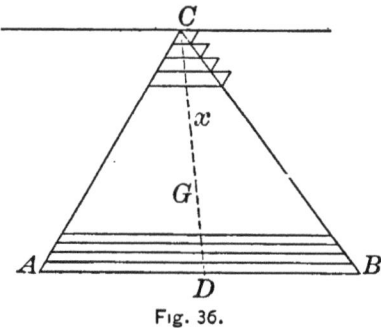

Fig. 36.

The areas of the small divisions of the triangle, beginning at the apex, are strictly as 1, 3, 5, etc. But no error will be introduced by assuming these areas, and therefore their masses, to be represented by the numbers 1, 2, 3, etc. For suppose the small triangular additions to be made to the several small areas as shown in the figure near C. The areas would then be strictly as 1, 2, 3, etc. Now since there is an infinite number of divisions n of the triangle, the base and altitude of the equal triangular additions are infinitesimal quantities. The area of each triangle is therefore an infinitesimally small quantity of

the *second order*.   The error made then by adding in an
infinite number of these excess triangles is only an infinitesimally small quantity of the *first order*, and this is
negligible in comparison with the finite area of the large
triangle.

If m is the mass of the first small area, the series of
masses will be represented by the numbers

$$m, \; 2m, \; 3m, \; \ldots \; nm.$$

Let the plane of reference be drawn through C perpendicular to CD.   Then the products mx will be represented
by the series

$$1^2 m, \; 2^2 m, \; 3^2 m, \; \ldots \; n^2 m.$$

Hence CG or $X = \dfrac{\Sigma mx}{\Sigma m} = \dfrac{m \, (1^2 + 2^2 + 3^2 + \, \ldots \, n^2)}{m \, (1 + 2 + 3 + \, \ldots \, n \,)}.$

Apply to the summing of the two series the formula of
Art. 15, $s = \dfrac{n^{m+1}}{m+1}$, in which m is the exponent.

Then $1^2 + 2^2 + 3^2 + \, \ldots \, n^2 = \dfrac{n^{2+1}}{2+1} = \dfrac{n^3}{3}.$

$1 + 2 + 3 + \, \ldots \, n = \dfrac{n^{1+1}}{1+1} = \dfrac{n^2}{2}.$

Hence $X = m\dfrac{n^3}{3} \div m\dfrac{n^2}{2} = \dfrac{2}{3}n.$

But since n is the number of equal divisions of CD,
$\dfrac{2}{3}n$ equals $\dfrac{2}{3}CD$.

PROBLEMS.

1.   Show that the centre of inertia of three equal weights placed
at the three corners of a triangle corresponds with the centre of
inertia of the triangle itself.

2.   Find the centre of inertia of a square from which one section
made by the two diagonals has been removed.

3. A uniform circular plate has a circular hole cut in it with a diameter equal to the radius of the plate, the two circles being tangent. Find the centre of inertia of the remainder.

4. A square is described on the base of an isosceles triangle. What is the ratio of the altitude of the triangle to its base when the centre of inertia of the whole figure is at the middle point of the base?

5. If two triangles have the same base and equal altitudes, show that the distance between their centres of inertia is one-third the distance between their vertices.

64. Centre of Inertia of a Pyramid. — The pyramid may have any base. Conceive it to be divided into small sections by a very large number n of equidistant planes drawn parallel to the base. Let g be the centre of inertia of the section at the base, and draw Cg (Fig. 37). Since the sections of the pyramid are similar figures, the line Cg passes through the centre of inertia of all these elementary areas. The centre of inertia of the whole pyramid must therefore lie on this same line. Let it be at G.

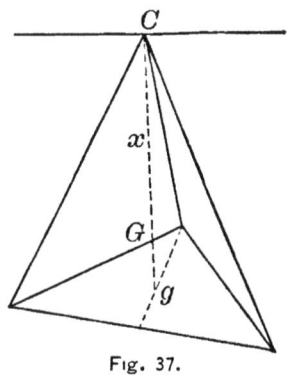

Fig. 37.

To find CG suppose a plane of reference drawn through C perpendicular to Cg. Then, since the elementary areas of equal thickness and density are similar figures, these areas are proportional to the squares of their homologous dimensions. The homologous dimensions are proportional to their distances from C, that is, to $1, 2, 3, \ldots n$.

If now the mass of the first section at C be represented by m, then the masses of all the sections may be represented by the series

$$1^2m, \ 2^2m, \ 3^2m, \ \ldots \ n^2m.$$

The products of these masses by their respective distances from the plane of reference will be the series

$$1^3m, \quad 2^3m, \quad 3^3m, \quad . \quad . \quad . \quad n^3m.$$

Hence $CG = X = \dfrac{\Sigma mx}{\Sigma m} = \dfrac{n^4}{4}m \div \dfrac{n^3}{3}m = \dfrac{3}{4}n.$

The centre of inertia of the pyramid is therefore on the line drawn from the apex to the centre of inertia of the base, and three-fourths of the distance from the apex down.

65. Moment of Inertia (A. and B., 56 ; L., 46.)— When a body rotates every point of it describes a circle around a line called the axis of rotation. Every point of the body then has its own velocity in its circle of rotation, and this velocity is proportional to the distance of the particle from the axis of rotation. Hence " to express the, speed with which a body rotates, it is sufficient to give the velocity of any one point together with its distance from the axis." The distance chosen to express the velocity of rotation, or the *angular velocity*, as this is called, is *unity*. Hence the angular velocity of a body is the linear velocity of a point situated at unit distance from the axis. This is represented by the Greek letter ω (Fig. 38), where Oa is unity.

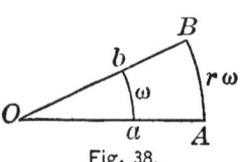

Fig. 38.

Since the linear velocity of any particle is proportional to its distance from the axis, the velocity of the particle m at A is $v = r\omega$, or $\omega = \dfrac{v}{r}$, where r equals OA.

Again if the velocity is uniform, a point at a distance of unity from the axis O describes a circumference 2π in one revolution of period T. Hence angular velocity is

$$\omega = \frac{2\pi}{T}. \quad \text{(Compare Art. 33.)}$$

The angular velocity represents the angle turned through per second by the whole body, as well as the distance travelled by a particle at unit distance from the axis in the same time. It must not be overlooked that the angle considered is always measured in circular measure.

Angular velocity may be variable. *Angular acceleration* is then the time-rate of change of angular velocity. Let a represent angular acceleration. Then

$$a = \frac{\omega - \omega_0}{t},$$

where ω_0 is the initial and ω the final angular velocity, the change being uniform throughout the period t.

If the initial velocity is zero, then

$$a = \frac{\omega}{t} = \frac{v}{rt}.$$

But $\frac{v}{t} = a$, the linear acceleration of a point at a distance r from the axis. Hence

$$a = \frac{a}{r}, \text{ or } a = ar ;$$

the angular acceleration is the linear acceleration of a particle situated at unit distance from the axis.

Let the particle m (Fig. 38) be at A, a distance r from the axis, and let its angular velocity be ω. Then its linear velocity is $r\omega$. But kinetic energy is half the product of the mass and the square of the velocity. The kinetic energy of the particle is therefore

$$w = \tfrac{1}{2} m r^2 \omega^2.$$

The kinetic energy of the entire rotating body is the sum of all such expressions as this last one, or

$$W = \tfrac{1}{2} \omega^2 \Sigma m r^2.$$

Since the angular velocity is the same for all points of the body, ω^2 is a constant and may be taken out from the sign of summation. The quantity Σmr^2 is called the *moment of inertia* of the body. It measures the importance of the body's inertia with respect to rotation. It is proportional to the kinetic energy of rotation. The work done upon a body to give it an angular velocity ω is therefore proportional to the moment of inertia of the body. The energy of rotation of a body whose angular velocity is ω depends not only upon its mass, but upon the manner in which that mass is disposed about the axis.

If the entire mass of the body is supposed collected at a distance k from the axis and so situated that the moment of inertia remains unchanged, then

$$I = \Sigma mr^2 = k^2 \Sigma m = M k^2.$$

The distance k is called the *radius of gyration*. The moment of inertia is usually represented by the letter I.

66. The Moment of Inertia and Angular Acceleration. — The moment of inertia may be defined as the moment of the couple required to produce unit angular acceleration. As shown in the last section, the linear acceleration a of a particle distant r from the axis is r times the angular acceleration or ra. Since force is the product of mass and acceleration,

$$f = mra.$$

The moment of this force about the axis is mr^2a. The acting distance is r, since the acceleration, and therefore the force, are directed tangentially to the circle in which m rotates. The total moment of all the forces producing the rotation of the entire body is therefore

$$Fb = a\ (mr^2 + m'r'^2 + m''r''^2 + \quad . \quad . \quad .) = a\,\Sigma mr^2.$$

If now a is unity, then the moment of the couple producing unit angular acceleration is Σmr^2, or the moment of inertia. Also

$$a = \frac{Fb}{\Sigma mr^2},$$

in which b is the lever arm at which the force F acts to produce the rotation.

PROBLEM.

The weight of a fly-wheel is M gms. Let it be considered entirely in the rim at a distance r from the centre. If a force of F dynes acts steadily upon the wheel at an arm of b cms., show that the angular velocity ω after t seconds from the commencement of the motion will be

$$\frac{Fbt}{Mr^2}.$$

67. The Moment of Inertia of a Thin Rod. — Let the axis of rotation be at right angles to the length of the rod through its middle point. Conceive each half of the rod to be divided into n equal divisions, each of mass m. Then the moment of inertia of the entire rod will be

$$I = 2m \left(1^2 + 2^2 + 3^2 + \ . \ . \ . \ n^2\right) = 2m\frac{n^3}{3}.$$

But the mass of the rod is $2mn$. Hence if M is the entire mass,

$$I = M\frac{n^2}{3}.$$

The length l of the rod is $2n$ or $n = \frac{l}{2}$, and $n^2 = \frac{l^2}{4}$. Substituting and

$$I = M\frac{l^2}{12}.$$

Since a rectangle may be conceived to be made up of an indefinite number of such parallel rods, if an axis be supposed to bisect two opposite sides of the rectangle, the moment of inertia for each elementary rod will be the expression above. If then m is the mass of each elementary rod or strip and a the length of the rod, or the bisected side of the rectangle,

$$I_a = \frac{a^2}{12} \Sigma m = M \frac{a^2}{12}.$$

If b is the other side of the rectangle, the moment of inertia for an axis bisecting the sides b is

$$I_b = M \frac{b^2}{12}.$$

The moment of inertia of a rectangle about an axis through its centre of figure and perpendicular to its plane is still larger.

68. The Moment of Inertia of a Circle. — (a) Let the axis be through the centre and at right angles to the plane of the circle. Conceive the circle to be made up of elementary rings of equal width and of radii

$$1,\ 2,\ 3,\ \quad .\quad .\quad .\quad n,$$

n being a very large number.

Let the mass of the inner ring be m and its radius unity. Then the masses of the several rings will be

$$m,\ 2m,\ 3m,\ \quad .\quad .\quad .\quad nm,$$

since they are of equal width and their circumferences are proportional to their radii. The density is of course assumed to be uniform. Hence we shall have for the moments of inertia of the successive rings

$$1m \cdot 1^2,\ 2m \cdot 2^2,\ 3m \cdot 3^2,\ \quad .\quad .\quad .\quad nm \cdot n^2 ;$$

and $\qquad I = m\,(1^3 + 2^3 + 3^3 + \quad .\quad .\quad .\quad n^3) = m\frac{n^4}{4}.$

But $\qquad M = m\,(1 + 2 + 3 + \quad .\quad .\quad .\quad n\,) = m\frac{n^2}{2}.$

Therefore $$I = M\frac{n^2}{2}.$$

Finally, since $n = r$, the radius of the circle,

$$I = M\frac{r^2}{2}.$$

Since a cylinder is made up of such rings, the moment of inertia of the cylinder about its axis also equals $M\frac{r^2}{2}$, in which M is the mass of the cylinder and r its radius.

(b) To find the moment of inertia of a cylindrical shell or ring about its axis, let r and r' be the external and internal radii; let M be the mass of the shell or ring and M' and M'' the masses of the cylinders of radii r and r' respectively. Then the moment of inertia of the ring is

$$I = M'\frac{r^2}{2} - M''\frac{r'^2}{2}.$$

Now $M' = \pi r^2 l d$ and $M'' = \pi r'^2 l d$, l being the length and d the density of the cylinder. Substitute and

$$I = \frac{\pi l d}{2}(r^4 - r'^4).$$

But M, the mass of the cylindrical shell, is $\pi l d (r^2 - r'^2)$.

Finally then $$I = \frac{M}{2}(r^2 + r'^2).$$

(c) The moment of inertia of a right cone, around the axis of figure, may be obtained in a similar way.

Divide the cone into thin circular laminæ by equidistant planes parallel to the base. The radii of these circles are

$$1, 2, 3, \quad . \quad . \quad . \quad n.$$

The masses are proportional to the squares of the radii. If the mass of the first one is m, the mass of the whole cone is

$$M = m\,(1^2 + 2^2 + 3^2 + \quad . \quad . \quad . \quad n^2) = m\,\frac{n^3}{3}.$$

The moments of inertia of the circular laminæ are

$$\frac{m}{2}, \; \frac{m}{2}2^2 \cdot 2^2, \; \frac{m}{2}3^2 \cdot 3^2, \quad . \quad . \quad . \quad \frac{m}{2}n^2 \cdot n^2.$$

Hence $I = \dfrac{m}{2}\,(1^4 + 2^4 + 3^4 + \quad . \quad . \quad . \quad n^4) = \dfrac{m}{2}\cdot\dfrac{n^5}{5}.$

'Substituting M for $\dfrac{mn^3}{3}$ and

$$I = \frac{3}{10}Mn^2 = \frac{3}{10}Mr^2,$$

if r is the radius of the base.

PROBLEMS.

1. Find the moment of inertia of a grindstone 1 metre in diameter and 10 cms. thick, density 2.14. The axis is through its centre and perpendicular to its plane.

2. Find the kinetic energy of the same stone when rotating 5 times in 6 seconds.

3. A homogeneous cylinder of mass m and radius a turns round a horizontal axis coinciding with the axis of the cylinder; a fine thread is wrapped round it, and a mass m' is attached to the end. Show that when the mass m' has descended through the height h the angular velocity of the cylinder, neglecting friction, is given by the equation $\omega^2 = \dfrac{4m'gh}{a^2(m + 2m')}.$

69. Moment of Inertia about a Parallel Axis. — Let G be the centre of inertia or centre of mass of the body; and suppose the moment of inertia of the body about the axis GY to be known, and let it be denoted by I_0. To find the moment of inertia about any parallel axis through A

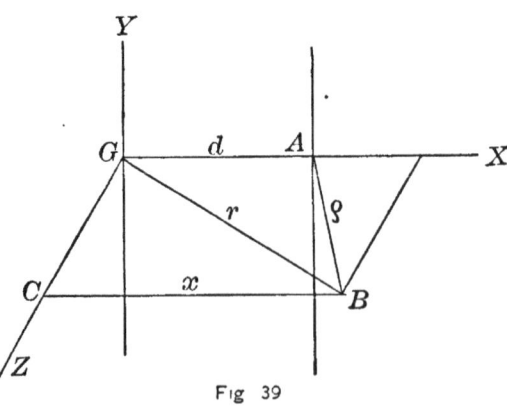

Fig 39

(Fig. 39), the distance between the two axes being d.

Let an element of the mass m be at B. GX, GY, GZ are rectangular axes. Then the figure $GCBX$ is a rectangle. In the triangle ABG

$$\rho^2 = r^2 + d^2 - 2rd \cos AGB.$$

Since the angle at C is a right angle,

$$r \cos AGB = x.$$

Therefore $\quad\quad \rho^2 = r^2 + d^2 - 2dx,$

and $\quad\quad m\rho^2 = mr^2 + md^2 - 2dmx.$

For every element of the mass a similar equation can be written. Therefore summing up for the entire body,

$$\Sigma m\rho^2 = \Sigma mr^2 + \Sigma md^2 - 2d\Sigma mx.$$

But by Art. 62 when the plane of reference, which is here the plane ZGY, passes through the centre of inertia, Σmx is zero. The last term of the above equation is then zero, or the sum of all the positive products mx is equal to the sum of all the negative ones. Hence if I is the moment of inertia about the axis through A,

$$I = I_0 + Md^2,$$

since Σm equals the entire mass of the body.

Therefore the moment of inertia about any axis exceeds the moment of inertia about a parallel axis through the centre of inertia by a quantity equal to the product of the mass of the body and the square of the distance between the two axes, or by a quantity equal to -the moment of inertia which the body would have about the axis through A if its mass were all aggregated at its centre of gravity G.

PROBLEM.

Two cylinders, of radius 1.007 cms. and mass 119.6 gms. each, are placed vertically on opposite ends of a bar suspended from its middle point so as to turn freely in a horizontal plane, the distance from the centre of the bar to the centre of each cylinder being 11.45

cms. Find the increase in the moment of inertia of the bar due to the addition of the cylinders.

70. The Ideal Simple Pendulum. — By an ideal simple pendulum is meant one in which the entire mass is supposed collected at a point and suspended by an inextensible thread without weight. Let the mass m be suspended from A (Fig. 40) by the thread of length l. In the position B the acceleration BG may be resolved into two components, the one in the direction of the length of the string, the ineffective component, and the other tangential, the effective component. The latter is

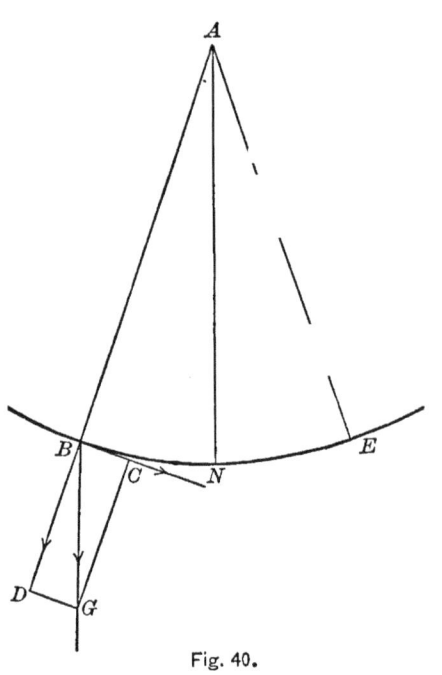

Fig. 40.

$$BC = f = g \sin \theta.$$

If the angle $BAN = \theta$ is small, $\sin \theta$ may be put equal to θ. Moreover, θ equals $\dfrac{BN}{AB}$ or $\dfrac{x}{l}$, the displacement BN being called x.

Hence
$$f = g\theta = g\frac{x}{l}.$$

The acceleration of the mass m at B is therefore proportional to its displacement x from the middle point of its path N. But this relation is characteristic of simple harmonic motion. The motion of the pendulum is therefore simple harmonic within the limits of the approximation that $\sin \theta = \theta$.

In simple harmonic motion

$$f = \frac{4\pi^2}{T'^2} x = \left(\frac{2\pi}{T}\right)^2 x.$$

Put
$$\left(\frac{2\pi}{T}\right)^2 = \mu.$$

Then
$$\frac{2\pi}{T} = \sqrt{\mu}, \quad f = \mu x, \text{ and } T = \frac{2\pi}{\sqrt{\mu}}.$$

This is the general formula for the period of oscillation in simple harmonic motion. But for the pendulum

$$f = \frac{g}{l} x.$$

For the simple pendulum therefore $\mu = \frac{g}{l}$.

Hence
$$T = 2\pi \sqrt{\frac{l}{g}}.$$

This is the period of a complete or double swing. For a single vibration

$$T = \pi \sqrt{\frac{l}{g}}.$$

The periods of pendulums of different lengths are proportional to the square roots of the lengths.

If T is one second

$$1 = \pi \sqrt{\frac{l}{g}},$$

or
$$g = \pi^2 l.$$

But $\pi = 3.14159$ and $\pi^2 = 9.869$.

Therefore $l = \frac{g}{\pi^2} = \frac{980}{9.869} = 99.3$ cms.

If the acceleration of gravity were 986.9 the seconds pendulum would be exactly one metre long.

The formula for the period of vibration is independent of the displacement x. Within the limits, therefore, of the approximation that the sine of an angle is equal to the angle itself, the vibrations are isochronous, or the period remains the same while the amplitude diminishes.

PROBLEMS.

1. Find the period of oscillation of a pendulum 6 metres long at a place where g is 980.

2. If the length of the seconds pendulum is 99.414 cms., what is the value of g?

3. A seconds pendulum is lengthened 1 per cent. How much does it lose a day?

4. A pendulum beating seconds at one place is carried to another station where it gains 10 seconds a day. Compare the values of gravity at the two places.

71. The Compound or Physical Pendulum. — Let the mass of a heavy pendulum, including its suspension, be M,

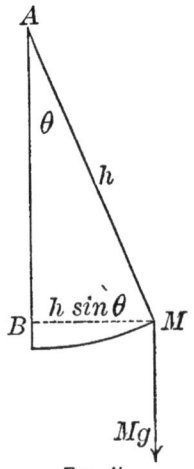

Fig. 41.

and let the centre of inertia of this mass be at a distance h from the axis of suspension A (Fig. 41). Then the total force of gravity, or the weight of the pendulum, is Mg, and the lever arm for the centre of rotation A is BM, or $h \sin \theta$.

The moment of the force producing rotation is therefore

$$Mgh \sin \theta.$$

The moment cf the force producing rotation is also (Art. 66) $a\Sigma mr^2$.

Hence $a\Sigma mr^2 = Mgh \sin \theta.$

From which

$$\frac{\Sigma mr^2}{Mh} = \frac{g \sin \theta}{a}.$$

The second member of this equation is the linear tangential acceleration divided by the angular acceleration.

This is a length (65) the same as linear tangential velocity divided by angular velocity. The equation may be considered then as a general formula for the length of the ideal simple pendulum which will oscillate in the same time as the real compound pendulum. That this is true may be demonstrated by supposing the mass M all collected at a point at a distance l from A, so situated that the time of oscillation as a simple pendulum is the same as that of the compound physical pendulum. Then the r for each particle becomes l, and h also becomes l for the aggregated mass. Hence

$$\frac{\Sigma m r^2}{Mh} = \frac{\Sigma m l^2}{Ml} = \frac{M l^2}{Ml} = l.$$

The length l is called the length of the equivalent simple pendulum, or the pendulum vibrating in the same time as the actual physical pendulum. It is found by dividing the moment of inertia of the pendulum about 'he axis of suspension by the product of its mass and the distance between its centre of mass and the axis of suspension.

By substitution in the formula of the last article, the period becomes

$$T = \pi \sqrt{\frac{\Sigma m r^2}{Mgh}}.$$

The numerator of the fraction under the radical sign is the moment of inertia about the axis around which the pendulum must oscillate, and the denominator is the *statical moment*, or the maximum moment of the couple producing rotation, when the lever arm is h or when

$$\theta = \frac{\pi}{2}.$$

If a line be drawn through the axis of suspension A and the centre of mass of the pendulum, and if the distance l be laid off on this line from A, its lower end will mark the point called the *centre of oscillation.* It is of course the point at which the whole mass must be collected so as to form the equivalent simple pendulum, and the body oscillates as if its whole mass were concentrated there. It is also called the *centre of percussion,* because, if the pendulum be started by a blow at this point, it will swing without any jar on its supports. A bat is an inverted pendulum, and drives the ball best without jarring the hands if it strikes the ball at the centre of percussion.

72. The Reversibility of the Pendulum. — Let Fig. 42 represent a physical pendulum consisting of a uniform

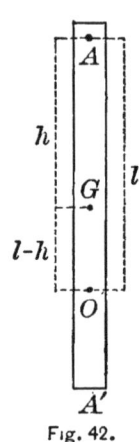

Fig. 42.

rectangle or bar. Also let A be the axis of suspension, G the centre of mass, and O the centre of oscillation.

Applying the principle of the moment of inertia about a parallel axis (Art. 69),

$$l = \frac{\Sigma mr^2}{Mh} = \frac{I_0 + Mh^2}{Mh}$$

for the axis through A.

Suppose the pendulum to be reversed and swung from an axis through O, and let l' be the length of the equivalent simple pendulum for this axis. Then

$$l' = \frac{\Sigma mr^2}{M(l-h)} = \frac{I_0 + M(l-h)^2}{M(l-h)}.$$

From the first equation

$$Mhl = I_0 + Mh^2,$$

or

$$I_0 = Mh(l-h).$$

Substitute this value in the second equation and

$$l' = \frac{Mh(l-h) + M(l-h)^2}{M(l-h)} = h + (l-h) = l,$$

or the length of the equivalent simple pendulum is the same whether the pendulum be swung from A or from O. If the length is the same the time of oscillation is also the same. This conclusion is of course independent of the form of the compound pendulum, or the arrangement of its mass about the axis. •

The length of the equivalent simple pendulum is then the distance between the two axes about which it swings in the same time.

73. Kater's Reversible Pendulum (V., 1, 238). — The reversible pendulum devised by Captain Kater utilizes the principle of the interchangeability of the centres of suspension and oscillation. A brass bar, carrying at one of its extremities a heavy lens-shaped weight, has fixed in it two knife-edges facing each other. One of these knife-edges is near one end of the bar and the other is next to the heavy weight at the other end, the weight being outside. Between the two is a smaller weight adjustable by means of a tangent screw. This weight is adjusted till the time of vibration of the pendulum is the same whichever knife-edge serves as the axis of suspension. The length l of the equivalent simple pendulum is then rigorously known and is equal to the distance between the two parallel knife-edges.

If its time of vibration is exactly determined by the method of coincidences, the length of the pendulum giving seconds in air can be determined. The mean of the results of Borda, Biot, and Peirce is

$$l = 993.92 \ mm.$$

at Paris at an elevation of 72 metres; this is for a vacuum and is doubtless correct to within $\frac{1}{100}$th of a millimetre. The value of g corresponding to this length is 980.96 at Paris.

74. Axis for Minimum Period of Oscillation. — If the pendulum is a uniform bar or rod, then for the axis of suspension at one end A (Fig. 43) the centre of oscillation is at O; and for an axis at the other end A', the centre of oscillation is at O'. The time of oscillation about these four axes is the same. If the axis passed through G the time of oscillation would be infinite, for then h is zero, and therefore l is infinite. The period of oscillation decreases as the axis is shifted from G toward O', and on further shifting to A the time is the same as at O'. If the axis could be further removed from G the period would continue to increase. It must therefore pass through a minimum between O' and A.

Fig. 43.

To find this point we have in Article 72

$$I_0 = Mh(l - h),$$

a constant, since I_0 is the moment of inertia about an axis through the centre of mass, perpendicular to the length of the bar.

Then $h(l - h)$ is a constant.

But when the product of two factors is a constant their sum is a minimum when they are equal to each other. Their sum

$$h + (l - h) = l,$$

and l is therefore least when

$$h = l - h,$$

or when

$$l = 2h.$$

The condition for the least period of oscillation is then that the axis of suspension and the centre of oscillation shall be equidistant from the centre of mass.

The length of the equivalent simple pendulum for this minimum period is twice the distance between the axis and the centre of mass.

CHAPTER V.

MECHANICS OF FLUIDS.

75. No Statical Friction in Fluids. — The term fluid applies both to liquids and gases. A *perfect fluid* would offer no resistance to a shearing stress ; but owing to viscosity no fluid is without shearing stress. The characteristic property of a fluid is to flow. Viscosity is a resistance to flow due to internal friction of the particles of the fluid against one another. A fluid possesses no rigidity, but is deformed by any force, however small. Ether, water, oil, molasses, Canada balsam, sealing-wax, shoemaker's wax, are examples of fluids of progressively greater and greater viscosity. Shoemaker's wax, which acts like a solid when struck with a blow, deports itself like a liquid because the deformation of a mass of it continues so long as the distorting force continues to act on it. It accommodates itself to an irregular, sinuous channel and flows slowly down it. Corks placed under a layer of shoemaker's wax on water gradually come through the wax to the surface. Bullets placed on top of the wax settle through it. Masses of wood in the wax slowly settle into the position of stable equilibrium which they immediately assume in water.

By statical friction is meant the friction existing between bodies relatively at rest. It acts tangential to the surfaces in contact. The absence of statical friction in fluids means

that all fluid pressure, when the fluid is at rest, is normal
to the surface of the fluid.   If this pressure were oblique
it could be resolved into a normal and a tangential component, and the latter would produce motion of the fluid.
Since action and reaction are opposite in direction the stress
between any two liquid surfaces in contact, or between a
liquid and the walls of the containing vessel, must be
normal to both.

When fluids are in motion there exists a tangential force
of friction.   Hence the middle of a stream moves faster
than the sides.   A glacier moves down its channel in the
same manner.   Tangential friction between two fluid surfaces is due to viscosity.

76. Pascal's Law. — In discussing the conditions of
equilibrium of liquids it is often useful to conceive small
portions to become solidified without change of density, or
to consider such portions as separately recognizable without
altering their relations to the surrounding mass.

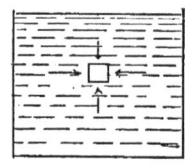
Fig. 44.

Conceive the small cube (Fig. 44) to
become solid without change of volume.
Then the pressure on each face of this
cube is the same, whichever way it is
turned, provided the fluid is at rest, and is
not acted on by any forces except those
applied to its surface.   If the forces on any pair of opposite
faces of the cube were unequal, their difference would
produce motion of the cube.   The dimensions of the cube
may become those of a material particle ; then the pressure
in all directions at the point must be the same.   If we
consider several such cubes touching one another along
any line, the pressure between them throughout the line
must be the same ; for there is no difference of pressure

on opposite sides of any one cube, and any difference between adjacent cubes in opposite directions would produce motion, which is contrary to the hypothesis of the liquid at rest. Hence, expressly eliminating the influence of gravity, the pressure throughout the mass of a liquid at rest is everywhere the same. This is known as Pascal's Principle.

It follows that an increase of pressure on any plane is transmitted to every other point in the fluid.

Pressure applied to any area of a confined fluid is transmitted to every other equal area, either of the fluid or the walls of the containing chamber, without diminution. This is the principle of the transmission of pressure. It is the principle applied in the Hydraulic Press, which is employed for compression purposes, for making lead pipe

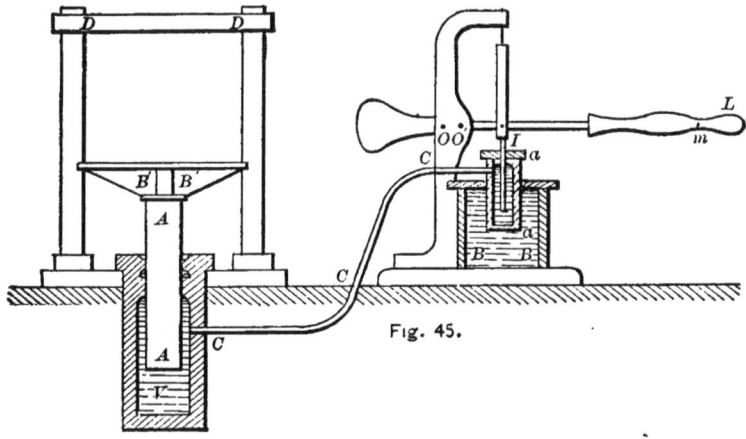

Fig. 45.

by forcing the lead through a die, and for the purpose of working cranes to lift heavy masses of metal, and the like. It consists fundamentally of two cylinders of different size, *aa* and *AA* (Fig. 45), connected by a pipe *CC*. The small cylinder *aa* is provided with an inlet and an outlet valve, not shown, similar to those of a force-pump. The plunger

in *a* is worked by the lever *L*. Water is thus drawn from
the surrounding reservoir, and is forced through the pipe
CC to the large cylinder *V*, where it acts on the plunger *A*.
The mechanical advantage of the machine is the ratio of
the cross-sections of the plungers *a* and *A*, so that a
pressure of *p* dynes on *a* balances $p\dfrac{A}{a}$ dynes on *A*. But
what is gained in power is lost in speed, for the plunger *a*
must travel $\dfrac{A}{a}$ times as far as the one in *A*. The work
done on *a* is thus equal to that done by *A* if friction is
neglected. There is no gain of energy, but the liquid
acts as a practically incompressible medium for the transmission of pressure.

**77. Pressure the same at all Points of a Horizontal
Plane.** — If the fluid is not weightless, Pascal's Principle
does not hold good, except for the same horizontal plane.
For in order that the cube above considered shall be in
equilibrium, the pressure on its lower surface must exceed
the pressure on its upper surface by the weight of the cube
itself. But the pressures at all points in a horizontal plane
still remain equal to one another. For consider a small
cylinder of the fluid with its axis horizontal and its ends
minute vertical planes. Then since the cylinder is at rest
the pressures on its ends are equal horizontal forces. But
the cylinder may be in any position in the plane. The
horizontal forces throughout the plane are then everywhere
the same. Moreover, the application of the reasoning
employed in the last article to any minute cube in the
plane would show that the forces acting on the cube reduced to a material particle are the same in all directions.
This holds true for all the particles throughout the plane;
so that the forces at all·points in the same horizontal plane

are the same in all directions. But while this is true for every horizontal plane, the value of the force changes from plane to plane because of the weight of the liquid.

78. The Free Surface of a Liquid Horizontal. — Consider a particle m at some point B (Fig. 46) of the free surface ABD, which we may suppose is not horizontal. The vertical force on m is mg, represented by BW. Resolve this into two componeuts, one normal and the other tangential to the sur-

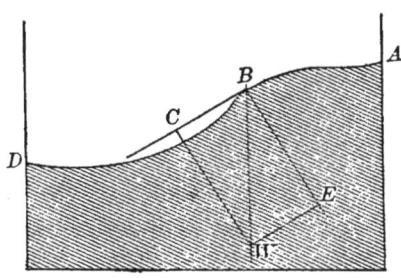

Fig. 46.

face at B. The latter component BC will cause the particle to move. But by hypothesis the liquid is at rest. Hence there can be no tangential component. But this tangential component disappears only when the surface is horizontal. Therefore, the free surface of a liquid is a horizontal surface, and the lines of force due to gravity are everywhere normal to it.

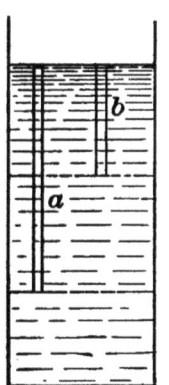

Fig. 47.

79. Pressure Varies directly as the Depth. — The pressure at every point in an incompressible fluid due to its weight is proportional to the depth beneath the surface. For consider two vertical cylinders of liquid with their upper ends in the surface of the liquid at rest (Fig. 47). Let the length of a be n times that of b. Then the weight of the one is n times as great as that of the other, and equilibrium can exist only when the upward pressure on the base of a is n times as great as on that of b; for the upward pressure is balanced by the weight of the column of liquid. The pressure on the plane at a depth n is therefore n times as great as at a depth unity, or the pressures are proportional to depths.

80. Two Liquids in Communicating Tubes. — Let two liquids of different density be placed in the two limbs of a bent tube (Fig. 48). Then their heights above the horizontal plane drawn through their surface of separation are inversely as their densities. For let their heights be h and h' and their densities d and d' respectively. Then since the pressure is everywhere the same on the horizontal plane of separation, the pressures due to the two columns of heights h and h' must also be the same. But the volumes for a column of unit cross-section are h and h', and volumes multiplied by densities give masses. Hence

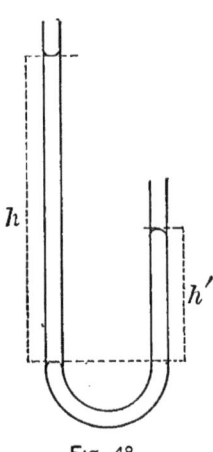

Fig. 48.

$$hdg = h'd'g,$$

or

$$\frac{h}{h'} = \frac{d'}{d}.$$

The heights are therefore inversely as the densities.

81. Back Pressure on a Discharging Vessel. — The vertical sides of a vessel containing a liquid sustain equal lateral pressures on the same level so long as the liquid is at rest.

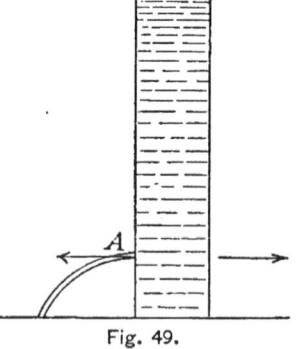

Fig. 49.

The pressures acting outward over the whole surface have therefore a resultant horizontally equal to zero. But let an opening be made at any point A (Fig. 49); the pressure inward at that point is removed by removing the reacting wall, while the outward pressure on the opposite wall remains uncompensated. Hence the entire vessel tends to move in the direction opposite to that of the stream. This is the principle of Barker's mill, of rockets, and of rotating fireworks.

It is perhaps better to apply to the discharging vessel the third law of motion. The momentum of the stream in one direction equals the impulse on the vessel in the other, or the action and the reaction are equal to each other. The same principle may be illustrated with air pressure by means of a contrivance similar to a Barker's mill screwed on the plate of an air-pump and covered by a receiver. The air is first exhausted. If it is then admitted at atmospheric pressure by means of the admit cock, the apparatus spins around with great rapidity on account of the uncompensated back pressure, or the reaction of the issuing jets of air on the walls of the tubes.

82. Total Pressure on any Immersed Surface. — Let a (Fig. 50) be any very small surface, the distance of

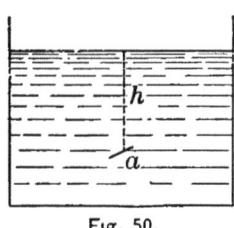

Fig. 50.

which from the free surface of the liquid is h. Since it is very small, the pressure sustained by it is independent of its position relative to the horizontal plane through it. This pressure is $ahdg$, in which d is the density of the liquid. The height h must be in cms., and a must be measured in square cms. The pressure will then be in dynes.

Let the area A be considered as composed of a large number of small elements, so that

$$A = a_1 + a_2 + a_3 + \quad . \quad . \quad .$$

Then the pressures on these areas will be

$$a_1 h_1 dg, \ a_2 h_2 dg, \ a_3 h_3 dg, \text{ etc.},$$

and the total pressure on A,

$$P = (a_1 h_1 + a_2 h_2 + a_3 h_3 + \quad . \quad . \quad . \quad) \, dg.$$

But by Art. 62 the quantities within the parenthesis may be put equal to AH, where H is the depth of the centre of inertia of the entire surface A, the plane of reference being the surface of the liquid. Therefore

$$P = AHdg.$$

This expression is the weight. of a prismatic or cylindrical column of the liquid, the base being the immersed surface A, and the height equal to the distance of its' centre of inertia below the free surface of the liquid. This is of course the pressure on one side of the immersed surface only. It applies equally well to the estimation of the pressure on any portion of the walls of the vessel or to retaining walls in general.

PROBLEMS.

1. Let a hollow cube be filled with water. The pressure on the sides is how many times the pressure on the bottom?

2. A spherical shell of radius r cms. is filled with water. Estimate the total hydrostatic pressure on the interior.

3. A retaining wall 2 metres wide and 50 metres long is inclined at an angle of 30° with the vertical. Find the total pressure of water in kilogrammes against it when the water rises to the top. Find also the horizontal pressure against it.

4. Find the pressure in grammes on the bottom and sides of a cubical vessel, 10 cms. on each side, filled with mercury. The density of mercury is 13.596.

83. The Centre of Pressure. — The centre of hydrostatic pressure on any immersed surface is the point of application of the resultant of all the elementary hydrostatic pressures against the elements of the surface. If the area is plane all these elementary pressures are parallel, and the problem consists in finding the resultant of a system of parallel forces. This problem we can treat here only in an elementary manner by means of a couple of examples. Thus, to find the centre of pressure on a rectangle with one side in the liquid surface. divide the rectangle (Fig. 51) into a very large number n of equal areas by lines parallel to the surface. Since the areas are equal the pressures on them are simply proportional to the depth, and the centre of pressure of each small area is on the line OG, which bisects all of them. Hence the centre of pressure of the rectangle is on this same line. Let it be at G. To find the distance OG.

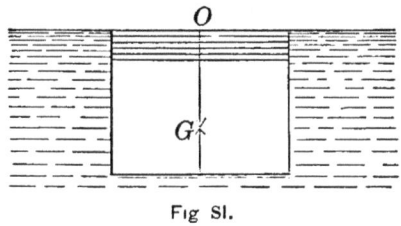

Fig 51.

The pressures on the several areas will be a constant multiplied by the integers 1, 2, 3, etc. This constant includes the area of each strip, the density of the liquid, and the cosine of the angle which the rectangle makes with a vertical plane.

Then the total pressure is

$$c\,(1 + 2 + 3 + \quad . \quad . \quad . \quad n) = c\frac{n^2}{2} = \Sigma p.$$

Then applying the general principle of Art. 62,

$$c\,(1^2 + 2^2 + 3^2 + \quad . \quad . \quad . \quad n^2) = c\frac{n^3}{3} = \Sigma px.$$

Hence $\quad X = OG = \dfrac{\Sigma px}{\Sigma p} = c\dfrac{n^3}{3} \div c\dfrac{n^2}{2} = \dfrac{2n}{3}.$

But n is the number of equal divisions or the total

depth of the rectangle. Hence the centre of pressure is
$\frac{2}{3}$ the depth of the rectangle.

84. Centre of Pressure on an Immersed Triangle. —

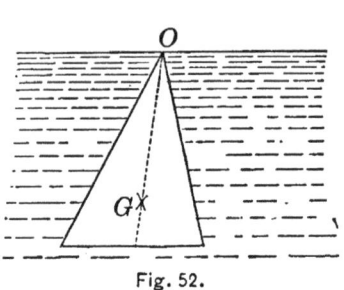

Fig. 52.

Let the apex be at the surface
and the base horizontal. Divide
the triangle (Fig. 52) into a large
number of sections n by equidistant lines parallel to the surface.
Then the areas of these sections
will be proportional to the depths
1, 2, 3, . . . n.

The pressures upon them are proportional jointly to the
areas and to their depths below the surface. We may
therefore write the total pressure in this case

$$c\,(1^2 + 2^2 + 3^2 + \quad . \quad . \quad . \quad . \quad . \quad n^2) = c\frac{n^3}{3} = \Sigma p.$$

Draw a line from O to the middle point of the base.
It will bisect all the small areas, and will therefore pass
through the centre of pressure of each one of them. The
centre of pressure on the triangle therefore lies on this
line. Let it be at G, at a distance X from O. Then the
products px may be written

$$c\,(1^3 + 2^3 + 3^3 + \quad . \quad . \quad . \quad . \quad n^3) = c\frac{n^4}{4} = \Sigma px.$$

Hence $\qquad X = OG = \dfrac{\Sigma px}{\Sigma p} = c\dfrac{n^4}{4} \div c\dfrac{n^3}{3} = \dfrac{3}{4}\,n.$

But n is the number of equal divisions of the line from
O to the middle point of the base. Hence the centre of
pressure is found by measuring three-fourths the depth
from the apex to the middle point of the base.

In a similar way, if the base were in the surface of the

liquid, it can be shown that the centre of pressure would be half way down from the middle point of the base to the apex.

PROBLEM.

A dam whose section is a right-angled triangle, 3 metres high, is built of stone of density 3. If the water reaches the top on the vertical side, what must be the breadth of the base with a factor of safety of ten, assuming that the wall may be treated as a rigid body?

85. The Principle of Archimedes. — When a body is immersed in any fluid it apparently loses weight. It is in reality partly supported by the fluid, or is subjected to an upward pressure equal to the weight of the fluid displaced. This follows at once if we consider that the body has replaced an equal volume of the fluid itself which was kept in equilibrium by an upward pressure equal to its own weight. The upward pressure on the immersed body is the same as that on the fluid which it replaces.

Let a cube of any heavy material be immersed in water (Fig. 53). The opposite lateral faces a and b will be equally pressed in opposite directions. The same will be true of the other pair of lateral faces. On d there will be a downward pressure equal to the column of water with base d and height nd. On the bottom c there will be an upward pressure equal to the weight of the column of water, whose base is c and height nc.

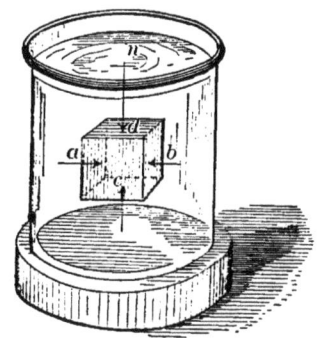

Fig. 53.

The resultant upward pressure on the solid is the difference of the pressures on the bottom and top of the cube, and this difference is the weight of the cube of water of the same dimensions as the solid.

Since the principle applies to gases as well as liquids, it may be stated generally as follows: A body immersed in a fluid is buoyed up by a force equal to the weight of the fluid displaced by it.

The *centre of buoyancy* is the name applied to the centre of mass of the displaced fluid.

When a body floats in a liquid it sinks to such a depth that the weight of the liquid displaced just equals its own weight. If the weight of the body is more than the weight of the liquid which it can displace, it will then sink; if less it will float. Thus a solid iron ball sinks in water, since its density is 7.8, or its own weight is 7.8 times as great as the water displaced by it. If, however, it is made in the form of a hollow ball, it may displace a mass of water equal in weight to its own.

On the contrary, iron cannot be made to sink in mercury, because the density of mercury is 13.596. Iron, therefore, floats on liquid mercury.

86. Density and Specific Gravity. — Density has already been defined as the mass in grammes contained in a cubic centimetre of volume.

Specific gravity is the ratio of the density of the body to that of another body taken as a standard. If s denotes the specific gravity of a body, d its density, and D the density of the standard, then
$$d = sD,$$
and
$$M = VsD.$$

If the density of the standard is unity, densities and specific gravities are numerically equal to each other. In the metric system the mass of a cubic centimetre of distilled water at 4° C. is one gramme, or the density of water is unity under standard conditions. The density and the specific gravity of any body in this system are, therefore, numerically identical.

If, however, the English gravitational system is employed then the density of water is about 62.4, since a cubic foot of distilled water at 4° C. contains a mass of 62.4 lbs.

87. Specific Gravity of Solids. — *a. Bodies heavier than water.* The density or specific gravity of a solid insoluble in water may be found by weighing the body first in air and then suspended in water. Its apparent loss of weight in water is, by the principle of Archimedes, the weight of the water displaced. Hence the quotient of the weight in air by the loss of weight in water is the density. If the water is at a temperature above (or below) the maximum, then the value found by the process just described must be multiplied by the density D of the water at the temperature at which the observation was made, or

$$d = sD.$$

b. Bodies lighter than water. Employ a sinker sufficient to make the body sink in water. Counterbalance with the body in the scale pan and the sinker suspended from the pan and immersed in the water. Transfer the body from the scale pan to the sinker in the water. The weight w', which must be added to the scale pan to restore the equilibrium, is the weight of the water displaced. It is not necessary to know the weight of the sinker. Then if w is the weight of the body in air, the apparent density is $\dfrac{w}{w'}$.

88. Density of Liquids inferred from Loss of Weight. — The density of liquids may be determined by means of the principle of Archimedes. A glass sinker is weighed in air, and then the loss of weight it sustains when immersed in water and in the liquid under examination is determined. Let these apparent losses be w and w'. Then the relative density of the liquid is $\dfrac{w'}{w}$, since its apparent loss of weight in the two cases is the weight of the same volume of the two liquids. If the water is not at 4° C. the result must be corrected as before.

89. General Theory of Hydrometers of Variable Immersion. — The approximate density of liquids may be conveniently determined by means of an hydrometer, which consists of a straight stem of glass, with a bulb at the bottom, and weighted so as to float to the proper depth in a vertical position. The graduation of hydrometers must be made experimentally. Those with equidistant divisions on the stem have their constants determined as follows : Let v be the volume immersed, *in units of the divisions of the stem*, when the hydrometer sinks to the zero of the scale. Then if d and d_1 are the densities of two liquids, and n, n_1 (Fig. 53a) the corresponding divisions of the scale to which the instrument sinks in them, the zero being near the top, the volumes immersed in the two cases are $v - n$ and $v - n_1$.

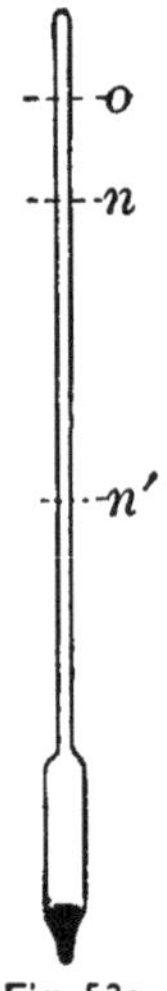

Hence
$$(v - n)\, d = (v - n_1)\, d_1,$$

since the masses of the liquids displaced are both equal to the mass of the hydrometer. Therefore
$$v = \frac{nd - n_1 d_1}{d - d_1}.$$

If one of the liquids is water and the hydrometer sinks in it to zero, then $n = 0$ and $d = 1$.

In that case
$$v = \frac{n_1 d_1}{d_1 - 1}.$$

Fig. 53a.

For any other liquid of density D, in which the hydrometer sinks to division N, the equation of equilibrium is

$$D (v - N) = v, \text{ or } D = \frac{v}{v - N}.$$

Since the density of water is unity its mass is numerically equal to its volume. The volume having been determined, a table can be made giving the densities corresponding to the various divisions of the scale.

90. Baumé's Hydrometer. — The two liquids are water and salt water, containing 15 per cent of salt, of density 1.11383 at 17°.8 C. The division to which the hydrometer sinks in the salt water is marked 15.

Then
$$v = \frac{15 \times 1.11383}{0.11383} = 146.78,$$

and
$$D = \frac{146.78}{146.78 - N}.$$

This formula gives the density corresponding to any division N.

For liquids lighter than water the zero of the scale is placed near the bottom of the straight stem. It is placed at the point to which the instrument sinks in a 10 per cent salt solution; the point to which it sinks in water is marked 10.

A general formula is obtained as before by reversing the signs of n and n_1, and making $n = 10$ for water. Then

$$v = \frac{n_1 d_1 - 10}{1 - d_1};$$

and if d_1 is the density of the 10 per cent salt solution for which $n_1 = 0$, then

$$v = \frac{10}{d_1 - 1}.$$

For other liquids

$$D = \frac{v + 10}{v + N}.$$

Gerlach obtained for v, 135.88.   Hence

$$D = \frac{145.88}{135.88 + N}.$$

91. Fundamental Phenomena of Capillary Action. —
Capillary action consists in the elevation or depression of
liquids along the walls of the vessel containing them, in the
ascent or depression of liquids between two plates very
close together, or in tubes of such small inner diameter
that they approach the dimensions of a hair; whence the
name capillarity, from *capillus*, a hair.

It is easy to determine that the free surface of a liquid
is not horizontal near the sides of the vessel containing it,
but presents a noticeable curvature.   When the liquid
wets the vessel, as water in glass, the surface is concave,
or the water rises along the glass ; on the other hand,
when the liquid does not adhere, as in the case of mercury
and glass, the surface is convex.

With the former conditions when small tubes, less than
two millimetres in diameter,
are supported in the liquid,
the liquid is perceptibly
higher in the tube than the
level surface without; with
the conditions determining
a convex surface the level
of the liquid within the

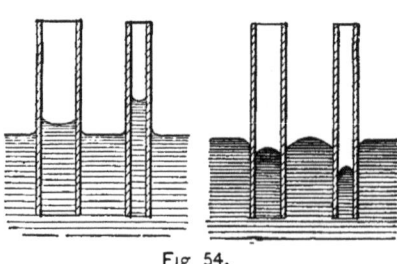

Fig 54.

tube is below that outside (Fig. 54).   This elevation or
depression of the liquid is inversely as the diameter of the

tube, provided this diameter does not exceed two millimetres.

The phenomena are independent of the pressure to
which the liquid is subjected.

They do not depend upon the thickness of the tube, or
in other words the action between the liquid and the tube
is limited to insensible distances.

The elevation or depression varies with the material of
the tube and the nature of the liquid. The elevation of
water in glass is greater than that of any other liquid,
being nearly three times as great as for sulphuric ether
and bisulphide of carbon.

Both the elevation and depression decrease with rise of
temperature. An elevation of water amounting to 132
mm. at 0° C. is reduced to 106 mm. at 100° C.

Capillary action explains the diffusion of liquids of slight
viscosity through porous septa, as well as their absorption
by porous bodies. Liquids which thus pass through porous
diaphragms are called *crystalloids;* while glutinous solutions of gum, starch, albumen, and the like, which do
not pass through porous septa, are called *colloids.* They
are viscous, diffuse slowly, and are indisposed to crystallize.
Physiologists attach great importance to this distinction,
inasmuch as it explains the interchange of liquids which
goes on through the tissues and vessels of the animal system, as well as the absorption of water by the spongioles
of roots.

Joseph Henry concluded that mercury passes through
lead by capillary action;[1] also that silver may penetrate
into the pores of copper when heated.

92. Law of Force assumed (A. and B., 90). — The
attraction of gravitation between masses of matter is much

---

[1] *Scientific Writings,* Vol. I., pp. 146, 228.

too small to account for capillary phenomena. But they can be explained if we assume an attraction between the molecules. The total expression for the stress between two molecules m and m' then becomes

$$F = C\,\frac{mm'}{r^2} + mm'f(r).$$

The first term expresses the attraction of gravitation; the second is the molecular attraction giving rise to capillary phenomena. All that is known about this function of r is that it is very large for insensible distances, that it diminishes very rapidly as r increases, and that it vanishes while r is still very small. The maximum value of r at which molecular action ceases, called ϵ, is estimated by Quincke to be 0.000005 cm. or 0.000002 inch. It is called the range or radius of molecular action. Within this distance the first term in the expression for the stress is quite negligible in comparison with the second.

The constant C in gravitation has a value of about $1/4000^2$, that is, a mass of nearly 4000 gms. placed at a distance of one centimetre from an equal mass would attract it with a force of one dyne.

93. Surface Tension (A. and B., 91; B., 201; Tait's Properties of Matter, 227). — If the molecules of a liquid are in equilibrium, then the conditions of the molecular balance in the interior of the liquid are different from those at the surface. At any point in the interior of the liquid, at a distance from the surface greater than ϵ, each molecule is attracted equally in all directions. But near or at the surface the attraction downward is not balanced by an equal attraction upward, and the molecules along the surface are therefore packed together so as to compose an enveloping film of thickness ϵ.

Consider a liquid bounded by a plane surface *mn* (Fig. 55), and let *m'n'* be a parallel plane at a distance ϵ below the surface.

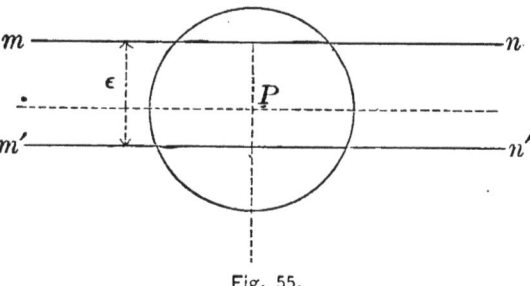

Fig. 55.

If we imagine any plane passed through a point in the mass of the liquid below *m'n'*, the normal pressure on this plane, due to molecular action, is independent of the direction of the plane with respect to the surface; for the number of molecules acting on the point is the same in every direction. If, however, the point is at *P*, nearer the surface than *m'n'*, about *P* as a centre and with a radius ϵ describe a sphere. Then the normal pressure on the plane through *P* perpendicular to *m'n'* is greater than when the plane is parallel to *m'n'*; for the upward attraction on the point varies from a maximum at *m'n'* to zero at the surface, since as *P* rises the upper half of the sphere described about *P* contains a diminishing number of molecules. From this inequality of pressure there results a stress or tension which causes the surface to contract; and this tendency to contract means that the surface acts like a stretched membrane.

If the surface be enlarged by forcing molecules out along the plane through *P* normal to the surface, then work must be done upon them to transfer them from the interior against the force pressing the molecules together along the surface. In other words, an increase in the area of the surface means an increase in the potential energy of the liquid. But potential energy always tends to become a minimum. The surface, therefore, contracts to as small dimensions as the conditions allow.

The volume enclosed by a sphere is a maximum, or the surface itself is as small as possible. A mass of free liquid always tends therefore to assume the spherical form except as it is distorted by other forces. Drops of rain and dew are spherical because of surface tension and not because of gravity. So also when a stream of molten lead flows from a small orifice, the surface tension causes the detached masses to form into spheres as the stream breaks. If they rotate as they descend, they remain quite spherical and strike the water at the bottom of the shot-tower as shot.

An ingenious method of separating the perfect shot from the imperfect ones consists in causing all together to roll down a smooth inclined plane. Near the bottom is a transverse slit in the plane. The perfect shot acquire enough momentum to carry them safely across, while the imperfect ones hobble along and fall into the crevasse.

The tendency of liquid masses to assume a spherical form is best illustrated by means of oil relieved from the effect of gravity by immersion in a liquid of its own density.

A mixture of alcohol and water is made of the same density as olive oil. Masses of olive oil placed in this liquid will neither rise nor sink, but will assume a globular form. If the limiting conditions imposed upon them do not permit them to assume the globular form, they will in every case assume interesting geometrical forms having the smallest superficial area under the given conditions. If, for example, a circular iron ring be immersed in a large mass of the floating oil, and some of the oil be then removed by means of a pipette, the remaining mass will take the form of a double convex lens.

94. Further Illustrations of Surface Tension. — Float two bits of wood on water parallel to each other, and a few millimetres apart. Let a drop of alcohol fall on the water between them, and they will suddenly fly apart. The surface tension of alcohol is less than that of water. The effect of the alcohol is therefore to weaken the film between the bits of wood. The parts of the film are thereby separated and carry the wood with them.

Place a thin layer of water on a piece of clean glass, and let a small drop of colored alcohol fall on it.

The weak spot made by the alcohol causes the film to break, while the tension about it draws the water away, leaving the alcohol surrounded by a dry area.

Make a ring of stout wire three or four inches in diameter (Fig. 56), with a handle. Tie to this a loop of thread so that the loop may hang near the middle of the ring. Dip the ring into a good soap solution containing glycerine, and obtain a plane film. T h e thread will float in it. Break the film inside the loop with a warm pointed wire,

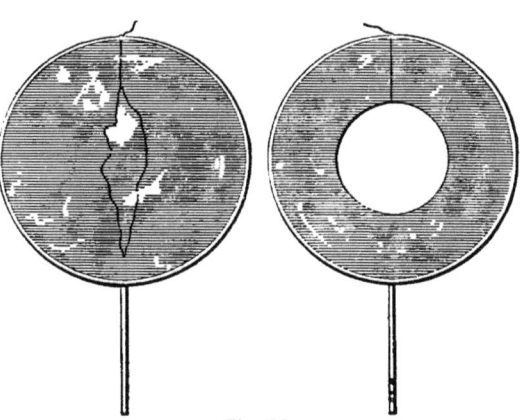

Fig. 56.

and the loop will spring out into a circle. The tension of the film attached to the thread pulls it out equally in all directions. By tilting the ring from side to side the circle may be made to float about on the film.

A small bit of camphor gum placed on warm water, perfectly free from any oily film, will execute rapid and irregular gyrations and movements across the surface. The camphor dissolves unequally at different points, and thus produces an unequal weakening of the surface tension of the water in different directions.

95. Energy and Surface Tension (A. and B., 93). — If we call the loss of potential energy, due to a diminution in the surface of one unit, the *surface energy* per unit area, it can be shown that this is numerically equal to the surface tension for unit width of the film. Let a liquid film be stretched on a frame *BCD* with the light rod *A* movable (Fig. 57). Let the length of the rod to which the film is attached, that is, the distance between *B* and *D*, be *a*, and let the rod be drawn toward *C* a distance *b*. Then the diminution in surface is *ab* units; and if *E* is the surface energy, the potential energy has decreased by an amount equal to *Eab*.

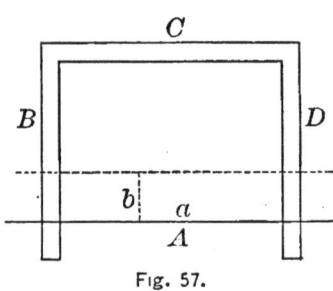

Fig. 57.

Further, if *T* is the surface tension per unit width of the film, the total surface tension lifting the rod is *Ta*. The distance moved is *b*. Hence the work done against gravity is *Tab*. This equals the loss in potential energy, or

$$Tab = Eab.$$

Therefore *T = E*, or the surface energy per unit area is equal to the surface tension per unit width.

If both sides of the film are taken into account the result is the same.

For a soap-film in air the surface tension is 27.45 dynes per centimetre width. Hence the surface energy is 27.45 ergs per square centimetre. For pure water and air the surface energy is 81 ergs per square centimetre.

96. Capillary Elevation or Depression explained by Surface Tension. — Let h (Fig. 58) be the mean elevation of the liquid in the tube above the liquid surface outside. The entire surface tension around the interior of the tube where the film is attached is $2\pi r T$, r being the radius of the tube. Let θ be the *angle of contact* which the film makes with the wall of the tube. Then the

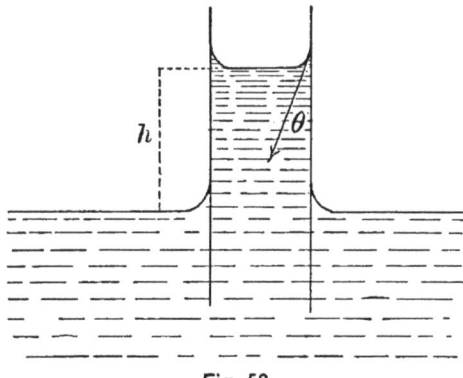

Fig. 58.

vertical component of the tension, pulling the liquid up and the tube down, is $2\pi r T \cos \theta$. This force is in equilibrium with the weight of the liquid column of height h. Let d be the density of the liquid. Then the weight of the column is $\pi r^2 h d g$. Consequently

$$2\pi r T \cos \theta = \pi r^2 h d g.$$

Therefore
$$h = \frac{2T \cos \theta}{rdg},$$

or the elevation is inversely proportional to the radius or diameter of the tube.

If the angle of contact is more than 90°, $\cos \theta$ is negative, and the elevation becomes a depression. If the liquid wets the tube there is an elevation; otherwise there is a depression. The curved surface of the liquid in the tube is called the *meniscus*.

For water the angle of contact with clean glass is supposed to be nearly or quite zero. Hence in this case

$$h = \frac{2T}{rdg}.$$

For two plates at a distance u from each other the total

tension for unit length along the plates is $2T$, and the
vertical component is $2T \cos \theta$. The weight of the liquid
column of cross-section u is $uhdg$. Hence

$$2T \cos \theta = uhdg,$$

or
$$h = \frac{2T \cos \theta}{udg}.$$

The elevation is therefore half as great as for a tube
whose *diameter* is u.

97. The Normal Pressure on a Curved Film. — A
stretched film with a curvature must always exhibit a
normal pressure directed
toward the concave side.

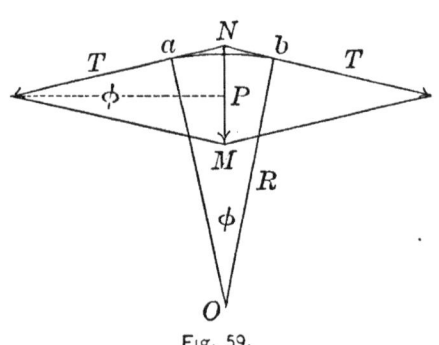

Fig. 59.

Let ab (Fig. 59) be a
small portion of the section
of a cylindrical film. T and
T represent the surface tension, stretching this film of
unit width perpendicular to
the plane of the paper, and
they are directed tangen-
tially at a and b. Complete the parallelogram and their
resultant is the diagonal NM. This is the normal pressure. Call it P'. Then

$$P' = 2T \sin \tfrac{1}{2}\phi.$$

But $\sin \tfrac{1}{2}\phi = \frac{ab}{2R}$, R being the radius of curvature of
the film or the radius of the cylinder. Therefore

$$P' = \frac{T}{R} ab.$$

If ab is unity, then the surface of the film considered is
one square unit, and the normal pressure per unit surface
is $P = \frac{T}{R}$, or T times the curvature.

Any other curved surface may always have its curvature expressed at any point by two principal radii of curvature, the planes of these curvatures being at right angles to each other. Let their radii be R and R_1. Then the normal pressure is the pressure due to the two curvatures conjointly, or

$$P = T\left(\frac{1}{R} + \frac{1}{R_1}\right).$$

If the film is plane, then both R and R_1 are infinite. If both sides of the film are free and the film is still curved, then the normal pressure is necessarily zero, and

$$\frac{1}{R} + \frac{1}{R_1} = 0.$$

This can be true only when $R = -R_1$, that is, the radii are numerically equal and the centres of curvature are on opposite sides of the film. Such a film is saddle-shaped, and it may easily be obtained by means of an oblong loop of wire, bent so that it does not lie in a plane.

For a soap bubble $P = 4\dfrac{T}{R}$, since there are two coneen-

tric spherical surfaces. Hence such a bubble always shrinks when the interior communicates with the outer air on account of the compression normally. The air inside a closed bubble must then be denser than the outer

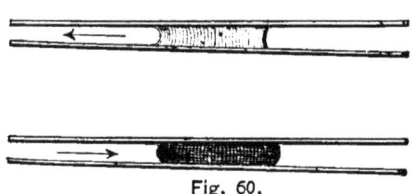

Fig. 60.

air, and minute vesicles of water filled with air are still heavier than the air displaced by them.

The normal pressure accounts for the motion of drops of liquid in conical capillary tubes. Thus a drop of water introduced into the larger end of a glass tube will move toward the smaller end (Fig. 60), while a globule of

mercury introduced into the smaller end will move toward
the larger.

98. **The Angle of Contact (A. and B., 95; B., 206).** —
Suppose the three dividing surfaces of three fluid substances in contact to meet along the line through O (Fig.

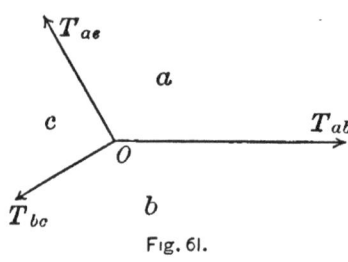

Fig. 61.

61), perpendicular to the plane
of the paper. Let T_{ab} be the surface tension between the media
a and b, T_{ac} that between media a
and c, and T_{bc} that between media b and c. Then if the three
tensions are in equilibrium, the
angles between them are con-
stant; for the three tensions may be represented by the
three sides of a triangle taken in order, and the angles between the three surfaces depend only upon the magnitude
of the three relative surface tensions.

But if T_{ab} is greater than the sum of T_{ac} and T_{bc}, then
the angle between T_{ac} and T_{bc} becomes zero, and the fluid
c spreads itself out in a thin sheet between a and b. This
is the case with oil between air and water. Let a be air,
b water, and c oil. Then

$$T_{ab} = 81 \quad \text{dynes.}$$
$$T_{ac} = 36.88 \quad \text{``}$$
$$T_{bc} = 20.56 \quad \text{``}$$

Hence $$T_{ab} > T_{ac} + T_{bc},$$

or $$81 > (20.56 + 36.88).$$

Hence a drop of oil on the surface of water cannot be
in equilibrium, but spreads itself out indefinitely thin between the air and water.

When two fluids a and b (Fig. 62) are in contact with a plane solid c, and their surface of separation makes an angle θ with the solid, the equation of equilibrium is

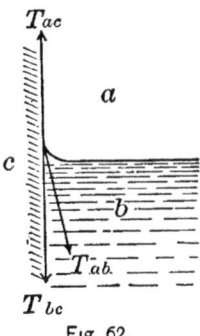

Fig 62.

$$T_{ac} = T_{bc} + T_{ab} \cos \theta.$$

But if T_{ac} be greater than the sum of T_{bc} and T_{ab}, the equation gives an impossible value for $\cos \theta$, the angle becomes evanescent, and the fluid b spreads out and wets the surface c. A drop of water will in this way spread out over the surface of a clean horizontal plate of glass; while a drop of mercury will gather itself together till the edges make a fixed angle with the plate.

99. Superficial Viscosity (D., 258). — Another property of liquid films, independent of surface tension, is their superficial viscosity. Surface tension is a constant stress in the bounding surface of a liquid, while superficial viscosity is a sort of surface friction which manifests itself only when something acts to rupture or otherwise disturb the surface film. A solution of saponine exhibits superficial viscosity to a marked degree. If a small magnetic needle be floated on the surface of it, the needle will remain in any position because the earth's magnetic directive force is unable to drag the liquid film around with the needle.

In most other liquids, when the needle turns, it carries with it the whole surface film, as may be shown by strewing on the surface lycopodium powder.

The viscosity of the surface film is, as a rule, much greater than the viscosity in the interior of the liquid.

Superficial viscosity holds a bubble on the surface of a liquid together, while the contraction of the surface due to surface tension tends to break it. Soapy water makes good bubbles, because while its surface tension is small its surface viscosity is large. A bubble rising through the liquid will raise a film at the surface which the surface tension cannot break.

Pure water has large surface tension, and relatively small superficial viscosity. Hence it does not froth.

Oil has small surface tension, but large surface viscosity or tenacity.

To this fact must be attributed the stilling of the sea when oil is poured on it. The new surface is relatively tenacious, and it is not readily broken into surf by the pressure of the waves from beneath.

" The superficial film of a liquid is thus seen to be a seat of energy, and to be physically different from the interior."

100. Air has Weight.— Aristotle attempted to determine whether air had weight by weighing a bladder collapsed and then inflated. Of course the change of buoyancy in the two cases offset the difference due to the weight of the air removed.

Galileo determined that water would not rise above 32 feet in the pumps of the Duke of Tuscany.

Since the invention of the air-pump it has been determined that air and hydrogen have the following weights:

1 litre of air weighs 1.2759 gms.,
1 litre of hydrogen weighs 0.08837 gms.,

both at a temperature of 0° C. and under a pressure of 10^6 dynes.

101. Atmospheric Pressure.— The total pressure of the atmosphere was first determined by Torricelli in 1643.

He took a glass tube, a little less than a metre long and

closed at one end, and filled it with mercury. Closing the upper end with the thumb he inverted the tube and placed the lower end under mercury contained in another vessel (Fig. 63). On removing the thumb the mercury fell in the tube, and came to rest at a height of about 76 cms. above the mercury surface in the outer vessel, leaving a vacuum in the tube above it, which has since been known as a Torricellian vacuum. It was rightly concluded by Torricelli that the mercury is sustained in the tube by the pressure of the atmosphere on the mercury surface external to the tube.

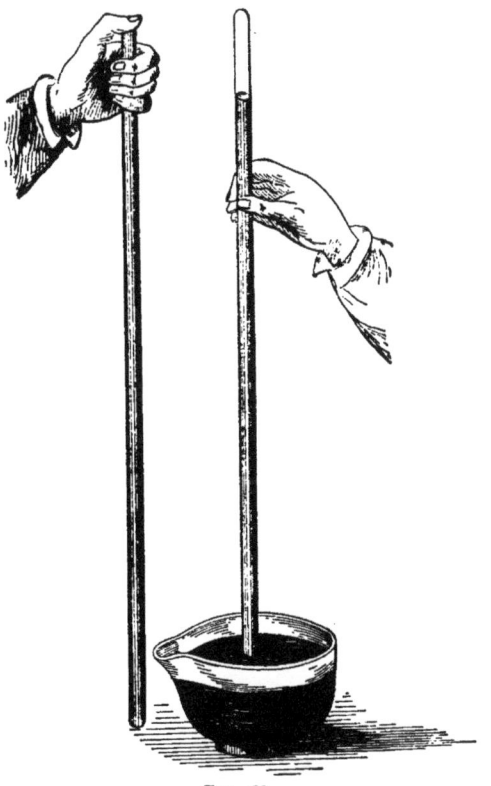

Fig 63.

Pascal performed two experiments to demonstrate that the column of mercury is supported by atmospheric pressure. In the first the mercury was replaced by lighter liquids, with the result that the height of the sustained column was always inversely proportional to the density of the liquid.

In the second experiment Pascal had the mercury column carried to the top of the Puy-de-Dôme, about 1,000 metres high. The pressure of the atmosphere being less on top of the mountain, it was anticipated that the mercury column would fall. A fall of nearly eight centimetres was observed.

The apparatus of Torricelli, when provided with a scale for the purpose of reading the height of the column of mercury, is called a *barometer*.

Atmospheric pressure on a square centimetre of surface is therefore the weight of the column of mercury one square centimetre in cross-section, and 76 centimetres in height at a temperature of 0° C. This is

$$76 \times 13.596 = 1033.3 \text{ gms.},$$

or $\qquad 1033.3 \times 980 = 1,012,630 \text{ dynes.}$

This is a little more than 10^6 dynes, a megadyne.

102. Height of the Homogeneous Atmosphere. — If the atmosphere were of the same density throughout its entire height as at the earth's surface this height in centimetres could be determined in terms of pressure, density, and gravity. It is often called the height of the homogeneous atmosphere. The pressure on one square centimetre would be

$$P = Hdg.$$

Whence $\qquad H = \dfrac{P}{dg}.$

P is the pressure of the atmosphere in dynes per square centimetre and d is the density of the air at 0° C. and a pressure of 76 cms. of mercury. Hence

$$H = \frac{1.0126 \times 10^6}{.001293 \times 980} = 7.99 \times 10^5 \text{ cms.}$$

For the same temperature at different elevations P varies directly as d. H therefore remains of the same value except for the change in g.

103. Boyle's Law (B., 190; A. and B., 141). — The law governing the compressibility of gases, *at a constant* *temperature*, was discovered by Robert Boyle in 1660. It is commonly known as Boyle's law, though it is sometimes ascribed to Mariotte. The law is as follows : The volume of a given mass of gas, at a constant temperature, is inversely as the pressure to which it is subjected. Expressed in symbols it is

$$\frac{v}{v'} = \frac{p'}{p}, \text{ or } pv = p'v'.$$

According to the law the product pv, at one temperature, is a constant.

Since volumes are inversely as densities, or

$$\frac{v}{v'} = \frac{d'}{d},$$

it follows that

$$\frac{d}{d'} = \frac{p}{p'},$$

or the densities are directly proportional to the pressures to which the gas is subjected.

Boyle's law is only approximately true. Such gases as sulphur dioxide, chlorine, and carbon dioxide, which are most easily liquefied by pressure, depart from the law most widely. Near the point of liquefaction the product pv is much smaller than Boyle's law requires.

Such gases as hydrogen, oxygen, and nitrogen follow the law most closely. These gases cannot be liquefied except by the combined means of great reduction of temperature and great pressure. They cannot be reduced to a liquid at ordinary temperatures by any pressure, however great.

They show a departure from Boyle's law different from that of the other class of gases. For every gas of this class, except hydrogen, a minimum value of pv has been found, beyond which, if the pressure is increased, the product pv increases. Thus if the product pv is taken as unity for air under a pressure of one atmosphere, at

77 atmospheres it becomes 0.9803; at 176 atmospheres, 1.0113; at 400 atmospheres, 1.1897. The minimum value of pv is at 77 atmospheres. If the experiments are made at a higher temperature the pressure at which the minimum value of pv occurs is greater and the agreement with the law is closer.

No minimum value of pv for hydrogen has been found, and this value must occur, if at all, for pressures less than one atmosphere. The value of pv for hydrogen is always greater than Boyle's law requires.

104. The Air-Pump. — The air-pump was invented by Otto von Guericke. Its action depends upon the elastic force of the gas by which it tends to expand indefinitely. Fig. 64 shows the essential parts of one of the best forms of the present. A piston P, with a valve S in it, works

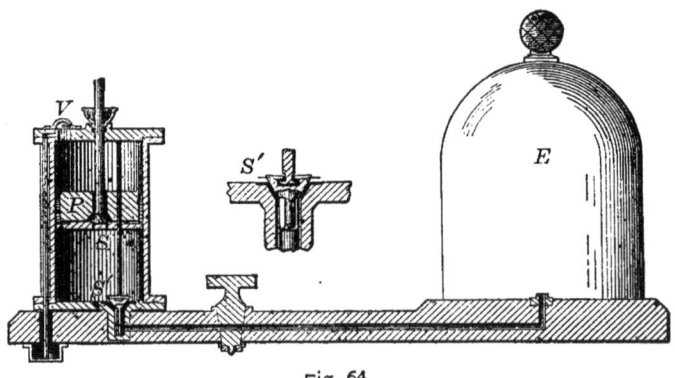

Fig. 64.

in a cylinder communicating with the air by a valve at its upper end opening outward, and with the receiver E by a valve S' at its lower end. When the piston ascends S is closed and S' is open. The gas expands and fills the cylinder. During the downward stroke S is open and S'

is closed. The gas thus escapes above the piston, and is forced into the open air when it is sufficiently compressed on the upward stroke to open the outward-opening valve. Let v be the volume of the receiver E, and c that of the pump cylinder. Let d and d_1 be the densities of the gas in the receiver before and after the first stroke and d_n the density of the gas remaining after the nth stroke. Then since the volume v becomes $v + c$ at the end of the first stroke, by Boyle's law

$$\frac{d_1}{d} = \frac{v}{v + c}.$$

For the second stroke

$$\frac{d_2}{d_1} = \frac{v}{v + c} \quad \text{again.}$$

Therefore after two strokes

$$\frac{d_2}{d} = \left(\frac{v}{v + c}\right)^2.$$

After n strokes

$$\frac{d_n}{d} = \left(\frac{v}{v + c}\right)^n = \left(\cfrac{1}{1 + \cfrac{c}{v}}\right)^n.$$

This expression can become zero only after an infinite number of strokes. But the limit of exhaustion by the mechanical air-pump is reached after a moderate number of strokes for several reasons. Among them are leaks at the valves and around the piston, untraversed space above and below the piston, and air absorbed by the oil used for lubrication.

105. Correction to the Weight of Bodies for Buoyancy of Air. — The apparent weight of a body is less than its real weight by the weight of the air which it displaces. But since this applies equally well both to the weights and

to the body weighed, a correction must be found depending upon their relative densities.

  Let x be the real mass of the body in grammes.
  Let w be the real mass of the weights.
  Let d be the density of the body.
  Let δ be the density of the weights.
  Let a be the density of the air.

Then the volume of the body is $\dfrac{x}{d}$.

The volume of the weights to counterbalance is $\dfrac{w}{\delta}$.

The masses of air displaced by the two are therefore
$\dfrac{x}{d}a$ and $\dfrac{w}{\delta}a$ respectively.

For equilibrium the equation is

$$x - \frac{a}{d}x = w - \frac{a}{\delta}w.$$

Whence $x = w\dfrac{1 - \dfrac{a}{\delta}}{1 - \dfrac{a}{d}} = w\left\{ 1 + a\left(\dfrac{1}{d} - \dfrac{1}{\delta}\right) \right\}$ nearly.

If d is greater than δ the correction is negative, that
is, the real weight is less than the apparent weight. If d is
less than δ, the correction is positive. The correction is
zero only when d and δ are equal to each other, or when the
body weighed and the weights are of the same density.

  For example, let a mass of sulphur, density 2, have an
apparent weight of 100 gms. when weighed with brass
weights of density 8.4. The correction is then positive,
and the true weight is

$$x = 100\left\{ 1 + \frac{1}{773}\left(\frac{1}{2} - \frac{1}{8.4}\right) \right\} = 100.049 \text{ gms.}$$

The correction is 0.049 gm.

106. Torricelli's Theorem for the Velocity of Efflux. — Let a small opening be made in the side of a vessel containing water, the depth of the orifice below the surface being h. Torricelli's formula for the velocity of efflux is

$$v^2 = 2gh.$$

This is the velocity which a heavy body would acquire in falling through the height h in a vacuum.

If we suppose a small mass m to issue from the orifice, an equal mass must have fallen some distance a_1 to take its place. Then another equal mass must have fallen a distance a_2, and so on through a series to the surface.

The total loss of potential energy is

$$mga_1 + mga_2 + mga_3 + \quad . \quad . \quad . \quad = mgh,$$

where h is the sum of a_1, a_2, a_3, etc.

If the loss in potential energy is all represented by the energy of motion acquired by the mass m, we may write

$$\tfrac{1}{2}mv^2 = mgh.$$

From this equation we obtain

$$v^2 = 2gh,$$

which is Torricelli's formula.

If a square centimetres is the area of the orifice, the number of cubic centimetres of water flowing out in t seconds should be avt, if Torricelli's formula is right. The quantity is always smaller than this because the effective area of the stream is not the area of the orifice, but the cross-section of the smallest part of the stream, or the *vena contracta*, which is about 62 per cent of the area of the orifice. The conical shape of the issuing jet is due to the lateral pressure on the water as it approaches the orifice. If a short cylindrical tube, whose length is two or three times its diameter, be placed in the opening so as to project outwards, the flow is increased to about 82 per cent of the theoretical amount.

107. Range of Jets. — Let ED (Fig. 65) be the side of a vessel of water, and let the surface be at E. Also let

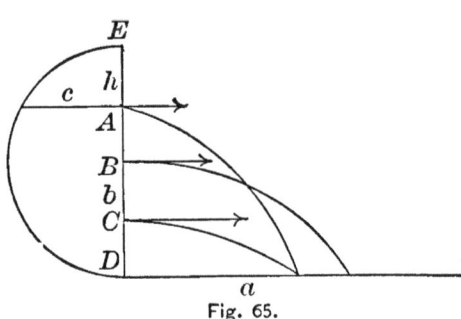

Fig. 65.

EA, AB, BC, and CD be equal to one another. Let v be the velocity of efflux from the orifice A. Then the range a of this s t r e a m will be

$$a = vt,$$

in which t is the time of falling to the horizontal plane through D. Also

$$b = \tfrac{1}{2}gt^2.$$

But if $\qquad v^2 = 2gh$, $a^2 = 2ght^2$, and $t^2 = \dfrac{a^2}{2gh}.$

Substitute this value of t in the equation for b and

$$b = \frac{a^2}{4h},$$

or $\qquad\qquad 4hb = a^2$ and $a = 2\sqrt{bh}.$

It follows that if h and b exchange values, the range will be the same, for their product bh is unaltered. This means that the range a of the jet from C is the same as of the one from A.

When the sum of two factors is a constant, their product is a maximum when the factors are equal to each other; the range a is therefore greatest for the opening B at the middle of the height, for then h and b are equal to each other, and their sum is the constant ED.

This may be shown in another way. On *ED* as a diameter describe a semicircle. Then the square of any half-chord equals the product of the two sections of the diameter; therefore

$$c^2 = bh \text{ or } c = \sqrt{bh}.$$

Hence $\sqrt{bh}$ will be a maximum when c is a maximum or at the point *B*. But the range a is $2\sqrt{bh}$, and is therefore greatest for the opening at *B*.

The form of the parabolic streams shows whether v has a value corresponding to Torricelli's formula. It is found to differ not more than one per cent from Torricelli's value $\sqrt{2gh}$.

108. The Common Siphon. — Let y (Fig. 66) be the height of the highest point of the siphon above the surface of the liquid in the vessel from which the discharge takes place; and let x be the height of the same point of the siphon above its open end or the lower surface of the liquid if the longer arm dips below the liquid. Let *H* be the height of a column of the liquid equal to atmospheric pressure. Then the pressure on one square centimetre of the highest cross-section of the siphon outwards is $d(H-y)$; and the pressure inwards on the same area is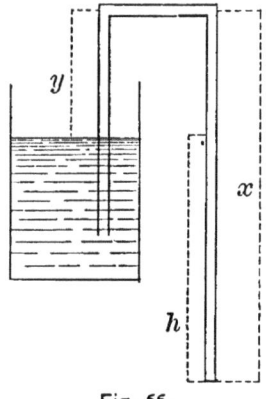

Fig. 66.

$d(H-x)$. The difference is the effective pressure, or head, producing the flow,

$$d(H-y) - d(H-x) = d(x-y) = dh.$$

The density of the liquid is represented by d. In the case of water d is unity and the head is h.

If y exceeds H, the liquid will not rise to the bend of the siphon by atmospheric pressure, and the flow ceases.

109. Mariotte's Flask. — Mariotte's flask (Fig. 67) is an arrangement to secure uniform velocity of efflux. The

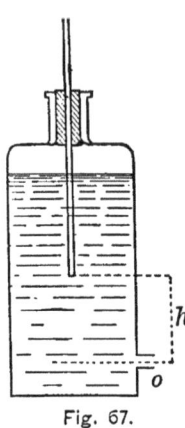

flask has a tubulure near the bottom as an outlet. A glass tube passes through an airtight stopper and extends down to within a distance h of the horizontal plane through the outlet o. The flow will then continue with a head h until the water descends to the lower end of the tube.

For the pressure at the lower end of the tube is the pressure of one atmosphere, the same as at the opening o. As water flows out air enters by the tube and takes its place, so that the effective pressure remains h. The pressure of the air in the flask and of the column of water above the lower end of the tube together constantly equal the pressure of the atmosphere.

Fig. 67.

PROBLEMS.

1. A uniform circular cylinder weighing 50 kilos. has a radius of 25 cms. and revolves without friction around a horizontal axis. A thread rolled around the cylinder carries at the free end a weight of one-half a kilogramme. Find how far the weight will descend from rest in three seconds.

2. A piece of silver and a piece of gold are suspended from the two ends of a balance beam with equal arms. The beam is in equilibrium when the silver is immersed in alcohol (sp. gr. 0.85), and the gold in nitric acid (sp. gr. 1.5); if the densities of the gold and silver are 19.3 and 10.5 respectively, what are their relative masses?

3. The length of the seconds pendulum at the equator being 990.93 mms., what is the corresponding value of g? What would be the value of g there if the earth did not rotate?

SOUND.

CHAPTER VI.

NATURE AND MOTION OF SOUND.

110. Sound and Hearing. — In all perceptions by the senses it is necessary to distinguish between the sensations themselves and the external cause of them — between the subjective and the objective aspects of the phenomenon. It is important to remember also that the objective causes of our sensations bear no *resemblance* to the sensations themselves. The external stimulus stands first in the series of energy-changes leading to a sensation, but it is not *like* the sensation. *Sound* should therefore be distinguished from *hearing* in the study of sound as a branch of Physics. All the external phenomena of sound may be present without the hearing ear.

All questions concerning sound come ultimately for decision to the organ of hearing; but in referring our sensations of sound to their external cause, we are only interpreting signs presented to consciousness, and drawing conclusions from them respecting outward phenomena. When this process, controlled by observation, experience, and trained reasoning, has led to the discovery of the physical facts constituting the foundation of sound, our investigations are largely transferred to the domain of mechanics (Lord Rayleigh).

111. The Source of Sound a Vibrating Body. — Very cursory examination serves to show that the source from which sound proceeds is always a vibrating body. " Sound and movement are so correlated that one is strong when the other is strong, one diminishes when the other diminishes, and the one stops when the other stops." [1]

Any regular succession of taps produces a musical sound. The element of regularity or periodicity is essential to make it musical. Otherwise it is mere noise. When a heavy toothed wheel is rotâted and a card is held against the teeth a musical sound of definite pitch is produced. So also the sound produced by a circular saw is musical at a distance where the highly discordant, irregular elements are eliminated.

If a goblet, partly filled with water, is set vibrating by drawing a bow across its edge, the tremors of the glass are communicated to the water and throw its surface into violent agitátion in four sectors, with intermediate areas of relative repose. All this ceases with the subsidence of the sound.

If a mounted tuning-fork is sounded and a light ball of pith or ivory, suspended by a thread, is brought in contact with one of the prongs at the back, it will be violently thrown away by the energetic vibrations.

If a minute globule of mercury is attached to a stretched wire by means of a little grease and lampblack, and is examined with a microscope, it will be found in motion, describing a line backward and forward, so long as the wire produces a musical sound.

A stout glass tube, several feet in length, may be made to emit a musical sound by grasping it by the middle and briskly rubbing one end with a moistened cloth. So

---

[1] *Blaserna on Sovnd*, p. **7.**

energetic are the longitudinal vibrations excited that it is not difficult to break the tube near the hand, and on the side opposite to the end rubbed, into many very narrow rings.

The vibrating body producing sound may be solid, liquid, or gaseous. Only the first and last are used in musical instruments, the first comprising all instruments employing strings, reeds, or bars, and the last including wind instruments of various sorts.

112. The Medium of Propagation. — Sound requires for transmission to the ear a continuous, ponderable, elastic medium; for the vibrations of a sonorous body cannot affect the organ of hearing without a medium of communication between them. If the vibrating body be isolated so that the required elastic medium does not extend to the source of vibrations, no sound will be perceived. This is somewhat imperfectly demonstrated, after the manner of Otto von Guericke, by suspending a bell by a thread in a receiver from which the air can be exhausted. The bell must not be allowed to communicate its vibrations directly to the pump plate. The sound becomes feebler as the exhaustion proceeds. Finally, if hydrogen be admitted and again exhausted, the sound will cease altogether, though the hammer may still be seen to strike the bell. When the medium about the bell is entirely removed it can no longer give up its energy to surrounding bodies, so as to set their molecules swinging with a to-and-fro motion; but its vibratory energy is converted into heat, in which form it can be propagated outwards by means of the ether. It vibrates longer and loses its energy more slowly when in an exhausted receiver than when it is beating the air. So a thin platinum wire or carbon filament continues to glow for several seconds after the electric current which heats it ceases to flow. But when surrounded by an atmosphere it cools very quickly, because it gives up its heat-energy to the surrounding gas, producing convection currents. The rarer the air at the source the feebler will be the sound.

The transmission of sound requires a medium both elastic and ponderable — elastic, because elasticity is the property by means of which the motion constituting sound is handed on from particle to particle; and ponderable, because sound is not transmitted through space exhausted of ordinary gross matter. The ethereal medium is not concerned in the transmission of sound.

113. Definition of Sound. — Sound may be defined as a vibratory movement excited in an elastic body, and transmitted to the ear by means of a continuous, elastic, ponderable medium. "Acoustics has for its object the study of those phenomena which may be perceived by the ear."

114. The Transmission of Sound (A. and B., 354). — The oscillatory motions of a sounding body are communicated to the air as the usual medium of transmission. The vibrations are said to be *longitudinal;* that is, in the direction of the sound-transmission. They are distinguished in this way from the *transverse* vibrations of water-waves, in which the motions of the particles are more or less nearly at right angles to the direction in which the waves are running. The vibrations of light are transverse.

The manner in which the motion is transmitted from particle to particle may be illustrated by means of a row of elastic balls lying in contact on two curved rails with the ends elevated. If one ball is allowed to roll down the groove and strike against the first one in line, the motion or impulse is handed on through the whole series, and the last ball moves up the incline. The *elasticity* of the balls explains the transfer of the motion through the series; the *energy* of the motion is independent of the elasticity of the conducting medium. It must all be supplied at the origin of the motion.

When the first ball strikes the second, compression takes place. The elasticity, called into activity by the distortion of the balls, tends to restore them to their unstrained form. The stress of elastic recovery is the same in both directions. The backward thrust brings the first ball to rest, while the forward one drives the second ball on against the third. The same operation is repeated between the second and third balls, and so on to the end of the series. But the last ball, not having any to which it can give up its motion, moves off up the incline.

To describe the motion by which sound is transmitted let *AB* (Fig. 68) represent an elastic cylinder, and let the layer *a* suffer a small displacement to the right. The

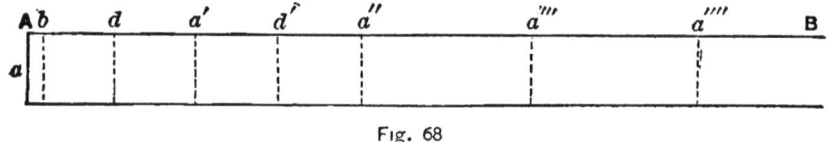

Fig. 68

effect of this displacement is that *a* approaches *b*, producing a condensation or crowding together of the particles. Therefore *b* is thrust forward, and the motion is communicated in the same manner from layer to layer through the cylinder.

If now *a* executes regular vibrations, its motions will

ultimately be communicated to all the other layers, because they are all tethered together as an elastic medium, and in time each layer of particles will be executing vibrations similar to those of a.

If the period of vibration of a is t, and the speed of transmission is v, then in one complete vibration of a the disturbance will travel a distance $s = vt$, or to a' in the figure. During two complete vibrations it will travel a distance $2s$ or to a''; in three periods to a''', and so on. The layer at a' therefore begins its first excursion as a begins its second; a'' begins its first as a' begins its second, and a its third, and so on. The layer at d, midway between a and a', begins its first vibration as a completes its first half-vibration; and it therefore moves *forward* while a moves *backward*. This related movement of particles in the cylinder constitutes a *wave*. While a is moving forward the particles near it constitute a *compression*; while it is moving backward they constitute a *rarefaction*. The distance aa', $a'a''$, etc., traversed by the disturbance during the period of a complete vibration of any one of the particles, is called a *wave-length*.

115. The Motion of the Particles and of the Wave. — The motion of the individual particles of the medium conveying sound is quite distinct from the motion of the sound-wave itself. This distinction is characteristic of all undulations transmitted through a medium of motion. A sound-wave is composed of a condensation followed by a rarefaction. In the former the particles of the medium have a forward motion in the direction in which sound is travelling; in the latter they have a backward motion, while, at the same time, both condensation and rarefaction are travelling steadily forward with a speed independent of that of the air particles. The independence of the two motions is aptly illustrated by a field of grain across which waves, excited by the wind, are coursing. No confusion between the two motions is here possible, because each stalk of grain is securely anchored to the ground, while the wave sweeps onward. Each head of grain in front of the crest of the wave is found to be rising, while all those behind the crest are at the same time falling. They all sway forward and backward, not *simultaneously*, but in *succession*, while the wave itself travels continuously forward.

In a sound-wave, therefore, the motion of the wave and the motion of the particles composing the wave are not identical ; a wave in air is in no sense a current ; the motion of the condensation and of the particles composing it are in the same direction ; while the motion of the rarefaction and of the particles in it are in opposite directions. The transmission of a wave is the transmission of energy, and not the transfer of the medium composing the wave. A series of particles along a line marking the progress of the wave are in successively different phases of their motion ; and the distance between two particles having the same phase in succession is a wave-length.

While any element of the medium merely oscillates about its position of rest, there is a continuous handing on or flow of the energy from point to point. In the case of a current, matter flows from one place to another, carrying the associated energy with it, so that there is a flow of both energy and matter.

A *wave-front* is the continuous locus of all points which are in the same phase of vibration, or of those portions of the medium which at the instant considered are equally and similarly distorted.

116. Experimental Determination of the Velocity of Sound in Air. — Since the motion of the particles of the medium is distinct from the motion of the sound-wave, the two kinds of motion admit of independent treatment and illustration. We shall consider first the velocity of sound in air and other media.

The usual determination of the velocity of sound is founded upon the measurement of the interval which elapses between the observation of some phenomenon first by sight and then by hearing. The observations have commonly been those of the flash and the report of a distant cannon. Since light is transmitted with such rapidity, the interval between the two observations may be regarded without sensible error as that which the sound actually requires to traverse the distance between the two stations. The earlier observations in the latter part of the seventeenth century and at the beginning of the eighteenth were not of sufficient value to report here. But beginning with those made by the French Academy of Sciences the following are some of the most trustworthy results :

1. Academy of Sciences, 1738 . . . . 332.00 metres.
2. Bureau des Longitudes, 1822 . . . . 331.00 "
3. Moll and Van Beck, 1823 . . . . 332.25 "
4. Stampfer and Myrbach, 1823 . . . . 332.44 ··
5. Bravais and Martins, 1844 . . . . 332.37 ··
6. Stone, 1871 . . . . 332.40 ··

The more precise measurements give a velocity of 332.4 metres at a temperature of 0° C. All of the above results have been reduced to this temperature.

In Stone's determination a cannon was fired, and two observers, three miles apart, gave signals by electricity on hearing the report. The eye observations were thus eliminated, as well as the perturbing effect of the violent disturbance near the source of the sound. Still two observers were necessary, and the reflex time in the two, required to perceive and to record the observation, may not have been the same.

From 1862 to 1866 Regnault made a long series of observations on the transmission of a shock or pulse through the water-pipes of Paris. The velocity was found to be somewhat less than in the open air. The disturbances at each station were recorded automatically by electricity. The principal conclusions may be summarized as follows: [1]

1. In a cylindrical pipe the intensity of the wave does not remain constant, but is enfeebled with the distance, and the more rapidly the smaller the pipe.

2. The velocity of sound diminishes at the same time as the intensity. In a conduit 1.1 metres in diameter the velocity was 334.16 for a distance of 749.1 metres, and 330.52 for a distance of 19,851.3 metres.

3. The velocity tends toward a limit, which is larger the larger the pipe. This fact is exhibited in the following table:

Diameter of conduit.	Velocity at zero.	Distance traversed.
0.108 m.	326.66	4055.9
0.216 "	328.18	6238.9
0.300 "	328.96	15240.0
1.100 "	330.52	19851.3

After all corrections had been made Regnault obtained for the limiting velocity 330.6 metres at 0° C.

4. The mode of production of the wave does not appear to have any sensible effect on the speed of propagation.

---

[1] Violle's *Cours de Physique*, Tome II., 67.

5. The speed of propagation in a gas is the same, whatever may be the pressure to which the gas is subjected.

117. Theoretical Determination of the Velocity of Sound (Phil. Trans., 1870, 277; Maxwell's Heat, 223). — Let $A_1 A_2$ (Fig. 69) be a tube of one square centimetre

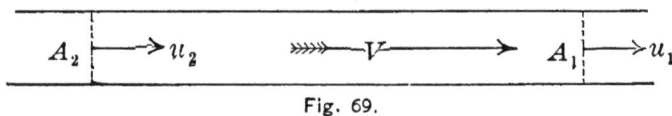

Fig. 69.

cross-sectional area and of indefinite length. Let A_1 and A_2 be two imaginary planes travelling with the velocity of sound V. Also let u_1, u_2 be the speed of the air particles at A_1, A_2; p_1, p_2 the corresponding pressures; and d_1, d_2 the densities. Then $V - u_1$ and $V - u_2$ are the velocities of the two planes *with respect to the medium;* and $(V - u_1) d_1$, $(V - u_2) d_2$ are the masses of air traversed by the two planes respectively in one second, since the cross-sectional area is one square centimetre. These masses are equal; for the two planes, travelling with the speed of the sound-wave, remain in the same relative position with respect to the condensation or rarefaction of the wave which they accompany, and there is therefore no accumulation or exhaustion of air going on between them during the motion. As much air streams in through A_1 as out through A_2. Moreover the mass of air traversed per second is the same as if the planes were travelling with a velocity V and there were no sound-wave. We may accordingly write

$$(V - u_1) d_1 = (V - u_2) d_2 = Vd = m . \quad . \quad . \quad (a)$$

The change of momentum of the mass m transferred in one second from one plane to the other is $m (u_2 - u_1)$.

But the rate of change of momentum is force, or, in this case, difference of pressure. Therefore

$$p_2 - p_1 = m (u_2 - u_1) \quad \cdots \quad (b)$$

From (a), $\quad u_1 = V - \dfrac{m}{d_1}; \ u_2 = V - \dfrac{m}{d_2}.$

Therefore $\quad u_2 - u_1 = m \left(\dfrac{1}{d_1} - \dfrac{1}{d_2} \right).$

Substituting in (b),

$$p_2 - p_1 = m^2 \left(\frac{1}{d_1} - \frac{1}{d_2} \right).$$

Let the volumes containing unit mass of air at densities d_1, d_2, and d be represented by s_1, s_2, and s. Then, since these volumes are the reciprocals of the corresponding densities, we have

$$p_2 - p_1 = m^2 (s_1 - s_2) = V^2 d^2 (s_1 - s_2).$$

Whence $\qquad V^2 = s^2 \dfrac{p_2 - p_1}{s_1 - s_2}.$

Let e be the coefficient of elasticity of the air. It is the quotient of the stress by the strain, or the quotient of the applied pressure by the voluminal compression produced. But the pressure producing the compression is $p_2 - p_1$, and the compression in volume is the diminution in volume $s_1 - s_2$ divided by the original volume s. Therefore

$$e = (p_2 - p_1) \div \frac{s_1 - s_2}{s} = s \frac{p_2 - p_1}{s_1 - s_2}.$$

Then $\qquad V^2 = s^2 \dfrac{p_2 - p_1}{s_1 - s_2} = se = \dfrac{e}{d},$

and $\qquad V = \sqrt{\dfrac{e}{d}}.$

118. Elasticity equals Pressure. — Let P and d represent the corresponding pressure and density of the air. Let the pressure be increased by a small quantity p and let the volume of unit mass of air be diminished in consequence by a small quantity s, the volume at pressure P being S. Then by Boyle's law, the temperature remaining constant,

$$P : P + p :: S - s : S.$$

By subtraction

$$p : P + p :: s : S,$$

or
$$\frac{p}{P+p} = \frac{s}{S}, \text{ the compression.}$$

Therefore the coefficient-of elasticity for *isothermal compression* is

$$e = p \div \frac{p}{P+p} = P + p.$$

If the disturbance is such as to make a relatively small change in density, p is negligible in comparison with P, and the coefficient of the elasticity of volume is equal to the pressure to which the gas is subjected. Then

$$V = \sqrt{\frac{P}{d}}.$$

But if the disturbances are violent, $p + P$ is no longer sensibly equal to P. The velocity for violent explosions is greater than for moderate sounds. Many observations confirm this conclusion. Captain Parry relates that in the arctic regions the report of a gun in artillery practice was often heard by a distant observer before the command to fire. Regnault found that the velocity of sound diminishes as the distance from the source increases. The same conclusion was reached by Stone.

Regnault also concluded that for musical sounds perceived by the ear the apparent velocity of acute sounds is sensibly less than of grave ones. But the observations are complicated by the fact that the sensation of hearing is excited more promptly by grave notes than by acute ones. The consequence is that when a sound travels through a great length of conduit it changes its quality or *timbre*.

It is a fact of common observation, however, that within moderate limits the velocity of sound is independent of pitch and loudness. If this were not so, then music, played by several instruments at a distance, would reach the listener out of time, and hence confused and discordant.

119. Newton's Form of the Equation for Velocity. — From Art. 102, $P = Hdg$, where H is the height of the homogeneous atmosphere.

Therefore $\quad \dfrac{P}{d} = gH$, and $V = \sqrt{\dfrac{P}{d}} = \sqrt{gH}.$

If a heavy body fall in a vacuum through a height H, the velocity attained is $v = \sqrt{2gH}.$

The speed of sound is therefore equal to the velocity acquired by a body falling in a vacuum through half the height of the homogeneous atmosphere. It was in this form that Newton announced the result of his investigation.

But $\quad\quad H = 7.99 \times 10^5$ cms. and $g = 980.$

Therefore $\quad V = \sqrt{980 \times 7.99 \times 10^5} = 27,972$ cms.

This is only 84 per cent of the observed velocity.

120. Corrections for Temperature (V., II, 52). — The effects on the velocity of sound in air, due to changes in temperature arising from two distinct causes, must be distinguished from each other. One change is that of the average temperature of the air through which the sound passes ; the other is that arising from compression and rarefaction in the two complementary portions of a sound-wave. The former may be considered as affecting only the pressure P ; the latter augments the elasticity independently of pressure.

First, consider the effect of a change of temperature upon pressure.

Let t represent the temperature and a the coefficient of expansion of a gas. The value of a is 0.003665.

The expression for velocity becomes then

$$V_t = V_0 \sqrt{1 + at} = V_0 \sqrt{1 + 0.003665t}.$$

The increase in velocity for one degree C. is therefore

$$V_0 \sqrt{1 + 0.003665} - V_0 = 0.00183 \; V_0.$$

Taking V_0 as 332.4 metres, the increase per degree C. is $332.4 \times 0.00183 = 0.608$ metre, or 23.9 inches.

Second. The value of the velocity of sound calculated from Newton's formula is $\frac{1}{6}$ less than that furnished by experiment. The cause of this disagreement was discovered by Laplace in 1816.

The coefficient of elasticity equals the pressure applied to the gas only under the condition applying to Boyle's law — that the temperature remain constant. In other words, it is the coefficient applicable to isothermal expansion or condensation when slow changes take place under a long-continued stress. But Laplace observed that by reason of the poor conductivity and radiating power of gases and the rapidity of the transmission of sound, the heat developed in any layer by the condensation could not immediately distribute itself throughout the entire mass; and that one should not, therefore, apply to it Boyle's law, which supposes a constant temperature. If, on the contrary, the heat remains entirely localized in the layer where it is produced, the phenomenon is subject to the formula of Poisson,

$$pv^\gamma = \text{constant},$$

instead of the formula of Boyle,

$$pv = \text{constant}.$$

γ is the ratio of the specific heat of a gas under constant pressure to its specific heat under a constant volume. The difference is that the coefficient of elasticity to be employed to bring the phenomenon under Poisson's formula is that corresponding to expansion or compression without the entrance or escape of heat. Such expansion or compression is called *adiabatic.* The heat effects are then all localized in the same masses of air where they are produced. This latter coefficient of elasticity is 1.41 times the other. The full formula then becomes

$$V_t = \sqrt{1.41 \; \frac{P\,(1+at)}{d}}.$$

121. Computation of the Velocity of Sound in Air. — The velocity of sound in dry air at 0° C. may be computed readily by the help of the formula.

Pressure P under standard conditions is 1,012,630 dynes per square centimetre (101).

The density d under the same standard conditions of 0° C. and 76 centimetres pressure of mercury is 0.001293.

$$\text{Therefore } V_0 = \sqrt{\frac{1.41 \times 1,012,630}{0.001293}} = 33,230 \text{ cms.}$$

122. Velocity of Sound in Water (V., II, 73). — The general formula

$$V = \sqrt{\frac{e}{d}}$$

is directly applicable; and since the compression of a liquid produces no appreciable heating effect,

$$e = \frac{\text{increase of pressure}}{\text{compression produced}} = \frac{mgH}{k}.$$

Here g is the acceleration of gravity,

     m the density of mercury,

     H the height of the normal barometric column,

     · k the coefficient of compressibility of the liquid or the compression.

A pressure of one atmosphere produces a compression k.

Then
$$V = \sqrt{\frac{mgH}{kd}},$$

d being the density of the liquid at the temperature of the observation.

At 4° C., k for water is 0.0000499, and its density is unity.

Therefore
$$V = \sqrt{\frac{980 \times 13.596 \times 76}{0.0000499}} = 142{,}500 \text{ cms.,}$$

or 1425 metres.

In 1827 Colladon and Sturm measured with much care the velocity of sound in the water of Lake Geneva between two boats anchored at a distance apart of 13,487 metres. The mean time required for the transmission of the sound of a bell struck under water was 9.4 seconds. This gives for the velocity at 8°.1 C., 1435 metres.

This is in very close agreement with the calculated value. The uncertainty relative to the value of k does not permit of a rigorous comparison between theory and experiment.

123. Velocity of Sound in Solids. — The same general formula, $V = \sqrt{\dfrac{e}{d}}$, is applicable to solids. Thus for copper, Young's modulus of elasticity, e, is 120×10^{10}, and its density is 8.8.

Therefore $V = \sqrt{\dfrac{120 \times 10^{10}}{8.8}} = 369,300$ cms., or 3,693 metres.

This is about 11.1 times as great as the velocity in air. Wertheim by an indirect experimental method found it 11.167 times as great.

For steel $e = 202 \times 10^{10}$, and d is 7.8.

Therefore $V = \sqrt{\dfrac{202 \times 10^{10}}{7.8}} = 508,400$ cms., or 5,084 metres.

124. Relations between Velocity, Wave-Length, Vibration-Frequency, and Period. — If n is the number of complete oscillations of an air particle per second, that is, the vibration-frequency; λ the wave-length or the distance the sound travels during a complete period of vibration T; then, since λ is the length of one wave and n such waves proceed from the source in one second, and extend over the distance V, it follows that

$$\lambda n = V, \text{ or } \lambda = \frac{V}{n}.$$

Also since the vibration-frequency is the inverse of the period, or $T = \frac{1}{n}$, we have

$$\lambda = VT.$$

125. Simple Harmonic Motion applied to Sound. — Let us next turn our attention to the motion of the medium. When a tuning-fork is set vibrating and is left to itself the intensity of the sound diminishes, but the pitch remains constant. Now pitch depends upon the vibration-frequency. The constancy of pitch therefore indicates constancy of vibration-frequency. The vibrations of the fork are isochronous, like those of the pendulum.

Unless an elastic body, like a tuning-fork or a stretched string, be very widely distorted, its periodic time, and therefore the pitch of the sound produced by it, are independent of the amplitude of vibration. Hooke's law of the proportionality of the forces of restitution to the distortion is a fundamental law of the vibratory motions which give rise to musical sounds. This means that in a tuning-fork, for example, the acceleration is proportional to the displacement.

We therefore conclude that the oscillations of the parts of musical instruments, as well as the swing of the air particles to which they give rise, may all be studied as simple harmonic motions.

126. Wave Motion as a Curve of Sines (A. and B., 356 ; Everett's Vibratory Motion and Sound, 46). — If the motions of the particles of air in sound are simple

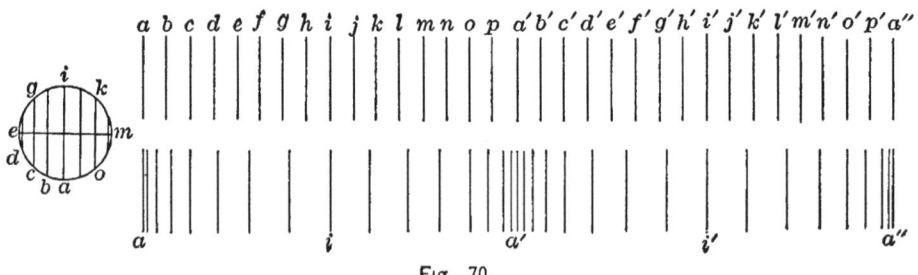

Fig. 70.

harmonic, then when a simple fundamental tone, without admixture with other higher tones, is transmitted through the air, the relative positions of the air particles along the line of transmission may be made out graphically by the help of the auxiliary circle employed in describing simple harmonic motion. Let the lines *a*, *b*, *c*, etc., Fig. 70, represent layers of particles, which in their state of equilibrium are at equal distances from one another. Suppose these layers of particles are executing simple harmonic motion of the same period and amplitude in a direction at right angles to their length, and that the phase of each layer is behind that of the preceding layer by a fixed amount depending on the distance between them; in the case figured it is one-sixteenth of a period.

Suppose further that the layer *a* is just passing through its position of equilibrium in the positive direction. We construct the circle of reference with a radius equal to the amplitude of motion for each layer. Then the projections of the points *a*, *b*, *c*, etc., on the diameter give the corresponding displacements of the layers by means of which the displaced position of each layer is found, as shown in the lower row of lines in the figure. The layers are crowded together as a condensation at *a*, *a'*, and *a''*, and separated as a rarefaction at *i* and *i'*. The distances *aa'*, *a'a''*, *ii'*, are all the same and are equal to one wave-length of the sound in air.

The vibrations of the particles in a sound wave are *longitudinal*, that is, in the direction of the motion of the wave; but for most purposes of graphical illustration it is easier to represent the displacement and the velocity of a particle by a line drawn at right angles to the direction of the motion, the length of the line indicating either the corresponding amount of displacement or the velocity, and its direction above or below indicating the·sign of the quantity. A sinusoidal curve may then be constructed to show either the displacement of successive particles or their velocities at a given instant. In this manner the curve of Fig. 71 was constructed by giving to each particle its proper positive or negative displacement, represented by its distance above or below the line OX. The points on the curve represent the displacements of successive particles at the instant when the particle a is passing its position of equilibrium in the positive direction.

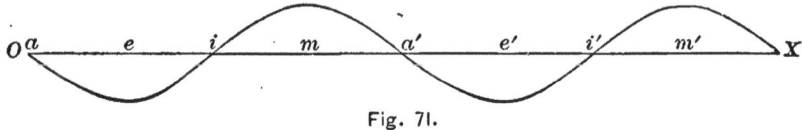

Fig. 71.

Such a curve may be drawn experimentally by causing a large tuning fork to inscribe its vibrations on smoked paper fastened round a drum, which can be rotated with a uniform angular motion while a light tracing point attached to the fork inscribes a sine curve.

To find the mathematical expression for this curve, let us suppose that to the first particle a is given a S.H.M. represented by

$$y = a \sin \frac{2\pi}{T} t \text{ (Art. 33)},$$

where y is the displacement. To obtain the displacement for a particle at a distance x from the origin, we must subtract from t the time required for the motion to go from the origin to x, which is x / v, if v is the velocity of propagation. Then for any particle we have

$$y = a \sin \frac{2\pi}{T} (t - \frac{x}{v}) = a \sin 2\pi (\frac{t}{T} - \frac{x}{\lambda}).$$

If we wish to know the position of successive particles at any instant when $t = nT$, or when a has made any number of complete vibrations, then

$$y = - a \sin 2\pi \frac{x}{\lambda}.$$

This equation shows that y is a *periodic* function of x, since the sine of an angle increasing with the time has regularly recurring values. The same value of y recurs with every increase of x equal to λ, the wave-length.

If various values are given to x the corresponding values of y will represent the displacements of the particles the distance of which from the origin is x. For $x = 0$, $y = 0$; for $x = \frac{1}{4}\lambda$, $y = -a$; for $x = \frac{1}{2}\lambda$, $y = 0$; for $x = \frac{3}{4}\lambda$, $y = a$; and for $x = \lambda$, $y = 0$. Laying off the several values of x along a straight line, and erecting perpendiculars equal to the corresponding values of y, the curve drawn through the extremities of all the ordinates is a curve of sines. Similarly the formula for the curve representing velocities is

$$y = \frac{2\pi a}{T} \cos \frac{2\pi x}{\lambda}.$$

From the two equations it is evident that the maximum velocity of a particle occurs simultaneously with minimum displacement and vice versa.

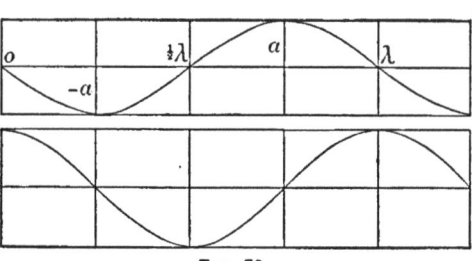

Fig. 72.

The ordinates of the upper curve in Fig. 72 represent the displacements of successive particles, and those of the lower curve the corresponding velocities at the same instant.

127. Composition of Simple Harmonic Motions in the Same Plane (A. and B., 359). — The curve of sines may be used to illustrate the composition of two or more wave motions in the same plane. If two systems of waves coexist in the same medium the displacement at any point will be the sum of the displacements due to the two systems taken separately. Hence the actual displacements may be found by taking the algebraic sum of the ordinates of the two displacement curves. If the two systems have the same period, then the resulting curve will be a simple sine curve of the same period; if the periods are not the same, the composite curve will not be a curve of sines.

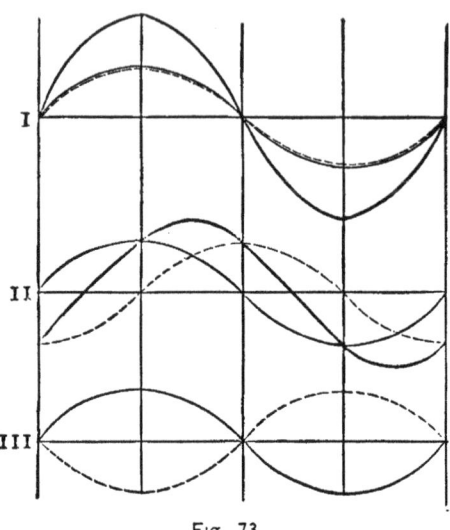

Fig 73.

In Fig. 73 the dotted and light lines represent the displacements due to two wave systems of the same period and amplitude. The heavy line represents the resulting displacement. In *I* the two systems have the same phase, and the resulting amplitude is double that of either component.   In *II* the phases differ by one-quarter, and in *III* by one-half a period.   In the last case the two motions completely annul each other. In Fig. 74 the periods of the two wave systems are as 1 to 2. The resulting curve is not a sinusoid, whether the component waves are in the same phase, as in the lower part of the figure, or not. The

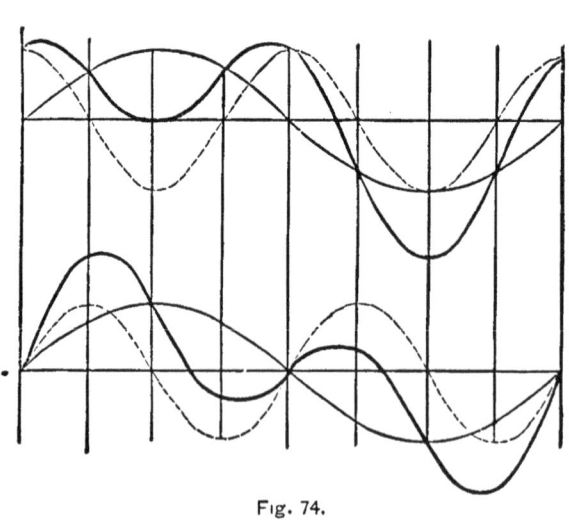

Fig. 74.

composite curve is periodic, however; that is, it has values of the ordinates recurring at equal time intervals.

128. Interference and Beats (V., II, 92). — The last topic illustrates the superposition of two wave systems in the same medium. In the case of two water-waves the total elevation or depression, relative to the primitive level, is at each point and at each instant equal to the algebraic sum of the displacements due to each system separately. If an elevation of the first system is superposed on an equal elevation of the second, the total height of the water above its primitive level will be double the elevation due to one system alone. If an elevation of the first system coincides with an equal and opposite depression of the second, the primitive level will not be modified.

In sound-waves condensations and rarefactions take the place of elevations and depressions, but two sound-wave systems modify each other in much the same manner as two systems of water-waves. Consider in particular two identical sources of sound *A* and *B*. It is clear that at every point of a plane, drawn perpendicular to *AB* at its middle point, the movements provoked by the two systems of waves at each instant will be concordant, or will reënforce each other. The velocity, as well as the displacement, will be greater than if there were only a single source of sound.

But in any other plane, parallel for example to *AB*, the velocity, and in consequence the intensity, will present a series of fixed maxima and minima. So two sounds emanating from two identical centres reënforce each other at certain points of space, and destroy each other at other points, or sound added to sound produces silence. This phenomenon is called *interference*.

When the two component sound-waves have nearly the same period the case deserves special attention. The reciprocal action results in periodic interference and *beats*.

For a few vibrations the periods may be regarded as the same, and the resulting vibration will be simple harmonic. But the more rapid vibration will gain on the other, thus changing the difference of phase on which the resultant depends. For when the two systems have the same phase they reënforce each other; when they have opposite phases they partially or wholly annul each other.

Let the two systems, of nearly the same amplitude, have vibration-frequencies m and n, $m - n$ being very small. Suppose the phases to agree initially. Then after an interval in seconds of $\dfrac{1}{2\,(m-n)}$ the two systems will be in exactly opposite phase, one system having gained half a wave-length on the other, and almost total extinction of motion and of sound will ensue.

After a further interval of $\dfrac{1}{2\,(m-n)}$ seconds, the system of shorter period will have gained a complete wave-length on the other, the two systems will again be in agreement in phase, and an increase of sound will result. This phenomenon, due to interference, is known as *beats*. The number of times per second that the two systems reënforce each other is equal to the difference in the vibration-frequencies of the two notes, or $m - n$. Thus if m is 103 and n is 100, then the first reënforcement will occur at the end of one-third of a second, a second one at two-thirds of a second, and a third at the end of the first second.

129. Experiments illustrating Interference of Sound-Waves. — Take two tuning-forks in unison, mounted on resonant boxes. Stick a small mass of wax to the prong of one of them. This increases its moment of inertia, and so increases its periodic time of vibration. If the two forks are now sounded together, the phenomenon of beats will be very pronounced.

Mount two organ pipes of the same pitch on a bellows, and sound together. If they are open pipes, a card gradually slipped over the open end of one of them will change its pitch enough to bring out strong beats. The same result may be produced by slowly sliding the finger across the embouchure of one pipe.

These two experiments illustrate interference of two sounds of slightly different pitch. The two sources are not identical. The following one will serve to illustrate interference from two identical sources, viz., the two prongs of the same fork.

Let *a* and *b* be the prongs of a diapason (Fig. 75). It is well known that the intensity of sound of a tuning-fork held freely in the hand and turned on its stem exhibits changes. As the two branches approach or recede from each other the movements communicated to the air are all the time opposing each other.

While the two branches, for example, approach each other, a condensation is produced between them, and at the same time two rarefactions start from the backs of the branches

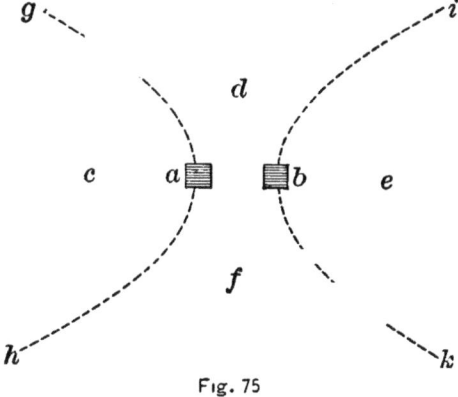

Fig. 75

a and *b*. When such a diapason is held before the ear, or is placed over a jar serving as a resonator, the sound is perceived to be strong in the regions *d* and *f*, less intense at *c* and *e*, and to disappear completely on the branches of the hyperbola *gah* and *ibk*. If then one turns the fork on its stem one perceives a succession of reënforcements and enfeeblements of sound. When the fork is in a position of least intensity of sound, the covering of one branch by a wooden or a pasteboard tube without touching serves to restore the sound to nearly maximum intensity.

130. To combine Two Simple Harmonic Motions at Right Angles. — In this case we seek the resultant motion arising from impressing simultaneously upon a particle two simple harmonic motions at right angles to each other. Each of the component motions may be regarded in the usual way as the apparent motion of a point moving uniformly around a circle. Let the radii of the two circles of reference be a and b. The periods of the two harmonic motions may have any ratio to each other, although only a few of the simpler ratios have been investigated.

Making use of the method adopted in Art. 33, the displacements in the two rectangular directions, if the periods are equal, will be

$$x = a \sin (\theta - \epsilon),$$
$$y = b \sin (\theta - \epsilon').$$

These two harmonic motions are entirely independent, but they are to be impressed upon the same particle. If $\epsilon' - \epsilon = \delta$, the difference of phase, we can express the two displacements under the form

$$x = a \sin (\theta + \delta),$$
$$y = b \sin \theta,$$

where δ is positive if the x component is in advance and negative if it is behind.

Finally, if n and m represent respectively the least numbers of complete oscillations executed by the moving point in the two directions in the same interval of time, then we have, as the most general form of the equations of displacement,

$$x = a \sin (n\theta + \delta),$$
$$y = b \sin m\theta.$$

In order to obtain the equation of the path of the moving point referred to rectangular axes, it is only necessary to eliminate the angle θ from the two equations and we have the "curve of impression as perceived by the eye."

131. To combine Two Simple Harmonic Motions of the Same Period at Right Angles. — For this case $n = m = 1$ and the displacements are

$$x = a \sin (\theta + \delta),$$
$$y = b \sin \theta.$$

Expanding the first equation

$$x = a (\sin \theta \cos \delta + \cos \theta \sin \delta).$$

From the second equation $\sin \theta = \frac{y}{b}$. Substitute in the equation for x and

$$x = \frac{a}{b} (y \cos \delta + \sqrt{b^2 - y^2} \sin \delta),$$

or

$$\left(x - \frac{ay}{b} \cos \delta\right)^2 = \frac{a^2}{b^2} (b^2 - y^2) \sin^2 \delta.$$

This is an equation of the second degree; and, since the curve returns into itself, or is a closed curve, it must be an ellipse. Consider four cases.

(1.) When $\delta = 0$ or there is no difference of phase between the two component motions. Then $\cos \delta = 1$; $\sin \delta = 0$. Consequently

$$x - \frac{ay}{b} = 0, \text{ or } y = \frac{b}{a} x.$$

This is the equation of a straight line through the origin, the inclination to the axis of x being $\tan^{-1}\left(\dfrac{b}{a}\right)$.

(2.)   When $\delta = \dfrac{\pi}{2}$ or the phase difference is a quarter of a period.   Then $\cos \delta = 0$; $\sin \delta = 1$.   Therefore

$$x^2 = \frac{a^2}{b^2}(b^2 - y^2),$$

or
$$\frac{x^2}{a^2} + \frac{y^2}{b^2} = 1.$$

This is the equation of an ellipse referred to its centre as origin and of semi-axes a and b.

(3.)   When $\delta = \pi$, or the phase difference is half a period.   Then $\cos \delta = -1$; $\sin \delta = 0$.   Therefore

$$x + \frac{a}{b}y = 0, \text{ or } y = -\frac{b}{a}x.$$

This is also an equation of a straight line through the origin, the inclination to the x axis being $\tan^{-1}\left(-\dfrac{b}{a}\right)$.

(4.)   When $\delta = \dfrac{3}{2}\pi$ or the difference of phase is $\frac{3}{4}$ of a period.   Then $\cos \delta = 0$; $\sin \delta = -1$.   Therefore

$$x^2 = \frac{a^2}{b^2}(b^2 - y^2).$$

This is the equation of the same ellipse as in case (2), but traced by the moving point in the opposite direction.

All the ellipses that can be obtained by varying δ will lie within a rectangle the sides of which are $2a$ and $2b$; if $x \cdot$ be made zero in the last equation above connecting x and y, y will equal $\pm b$; while if y be made zero, x will be $\pm a$. The ellipses are always tangent to the sides of the rectangle.

132. To combine Two Simple Harmonic Motions at Right Angles with Periods as One to Two. — If we assume the period of the x component to be one-half that of the y component, then $n = 2$, $m = 1$, and the two equations of displacement become

$$x = a \sin (2\theta + \delta),$$
$$y = b \sin \theta.$$

The elimination of θ from these equations gives in general an equation of the fourth degree, of which the three following cases are of most interest:

(1.) When $\delta = 0$ or π, $\sin \delta = 0$. Then

$$\sin \theta = \frac{y}{b} \; ; \; \frac{x}{a} = \sin 2\theta = 2 \sin \theta \cos \theta.$$

But if $\qquad \sin \theta = \frac{y}{b}$, $\cos \theta = \sqrt{\frac{b^2 - y^2}{b^2}}.$

Therefore $\qquad \frac{x}{a} = \frac{2y}{b} \sqrt{\frac{b^2 - y^2}{b^2}},$

or $\qquad x = \frac{2ay}{b^2} \sqrt{b^2 - y^2},$

or $\qquad b^4 x^2 = 4a^2 y^2 (b^2 - y^2).$

This is the equation of a lemniscate symmetrically placed on the y axis.

(2.) When $\delta = \frac{\pi}{2}$. Then $\cos \delta = 0$; $\sin \delta = 1$.

Therefore $\quad x = a (\sin 2\theta \cos \delta + \cos 2\theta \sin \delta),$

or $\qquad x = a \cos 2\theta = a (2 \cos^2 \theta - 1).$

Substitute now the value of $\cos \theta$ from (1) and after reduction

$$y^2 = \frac{b^2}{2a} (a - x).$$

This is the equation of a parabola symmetrically placed on the x axis, whose vertex is at a distance a from the origin, and focus at the distance $\frac{b^2}{8a}$ from the vertex.

(3.)　When $\delta = \dfrac{3}{2}\pi$.　Then $\cos\delta = 0$; $\sin\delta = -1$.

The equation then becomes

$$y^2 = \frac{b^2}{2a}(a+x),$$

denoting the same parabola reversed, its vertex being now turned in the direction of the negative axis of x.

133. Graphical Method of combining Two Simple Harmonic Motions at Right Angles. — Draw two concentric circles (Fig. 76) with radii proportional to the

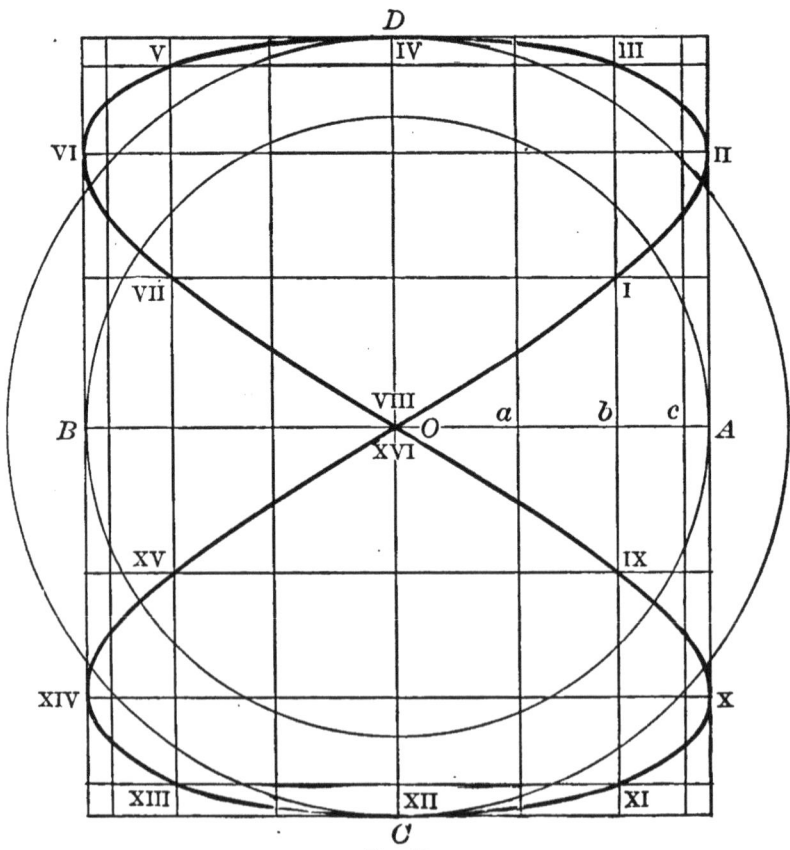

Fig. 76.

amplitudes, a and b, of the two harmonic motions, and through their common centre O draw the rectangular diameters AB, CD.

Divide each quadrant of both circles into the same number of equal parts; some multiple of four is usually most convenient. Through the points of division of the circle AB draw lines parallel to CD, and through the divisions of CD draw lines parallel to AB. The resulting rectangle of sides $2a$ and $2b$ will contain all the figures arising from any possible combination of two simple harmonic motions of commensurable periods; and the curves will, in general, be tangent to the sides of the rectangle. The centre of the circles corresponds to a phase difference of zero between the two components, that is, to $\delta = 0$; and it is taken as the starting point for tracing all curves of phase difference zero or π.

If, as in Fig. 76, the circles have been divided into sixteen equal parts, then each point of intersection on the diameter AB corresponds to a phase difference of $\frac{2\pi}{16}$ or $\frac{\pi}{8}$, that is, to one-sixteenth of a period. Hence if we start to trace a curve from a in the figure instead of from O, we shall produce the curve corresponding to a phase difference of $\frac{\pi}{8}$. This means that at the instant when the y component passes through AB in the positive direction, and the y displacement is therefore zero, the x component has already reached a in the positive direction, or is in advance of the y component by Oa or $\frac{\pi}{8}$. In like manner b corresponds to a phase difference of $\frac{\pi}{4}$, c to $\frac{3\pi}{8}$, and A to

$\frac{\pi}{2}$. Returning toward O, it will be seen that c also corresponds to a difference of phase of $\frac{5\pi}{8}$, b to $\frac{3\pi}{4}$, a to $\frac{7\pi}{8}$, and O to π, with larger values for points to the left of O.

Suppose now that we wish to trace the curve corresponding to the vibration-frequencies one to two, two for the horizontal and one for the vertical component, and with no difference of phase. Starting from O we count two points horizontally to the right and one up and reach I; again two to the right and one up for point II; and so continue, numbering the points in order until we pass through the starting point in the same direction as at first, being careful always to complete the motion in one dirce-tion before beginning the retrograde motion. An excellent cheek upon the accuracy of the location of the points is found in the fact that points equidistant from the axis of symmetry AB differ in number by eight in every case, that is, by half a vibration.

If now a smooth curve be traced through the points in order we see that, in accordance with Art. 22, the moving point, being subject to both motions, describes two spaces horizontally and one vertically in the same interval of time, and consequently passes through the corners of rectangles two spaces long and one space high in every case. The spaces themselves increase or decrease according to the simple harmonic law. Great diversity of figure may thus be obtained with successive differences of phase between the two component motions.

To combine two motions of frequencies two to three, we should simply count three spaces in one direction and two in the other and proceed in other respects as already described.

The curves obtained in this way are most beautifully verified experimentally by Blackburn's pendulum, with adjustable periods of vibration in two planes at right angles; or by Lissajous' optical method, in which a beam of light is successively reflected from mirrors on two tuning-forks, one vibrating horizontally and the other vertically.

134. The Principle of Huyghens (D., 110; P., 52; L., 229; A. and B., 356). — Let a (Fig. 77) be a centre of disturbance, and *mcn* the front of a spherical wave diverging from it. The radius of the wave increases with the velocity of sound, and the disturbance now at *mcn* will a moment later be at *m'dn'*. This single wave as it travels outward will disturb all the elements of the medium over which it passes. The disturbance of any one element of the medium may then be considered as the cause of the subsequent disturbance of all the other elements. The principle of Huyghens is that every point of the wave-surface *mn* becomes a new centre

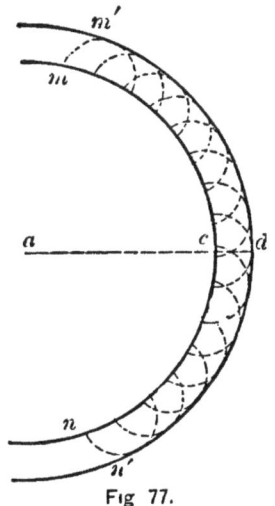

Fig 77.

of disturbance from which waves of sound are propagated outwards in the same manner as from the original centre; and the aggregate effect at any point outside this surface is the resultant of the combined action of all the secondary waves propagated from these new centres. The principle follows from the consideration that every particle on the wave-surface has the same oscillatory motion, except in point of amplitude, as the first particle disturbed; and it therefore stands in the same relation to adjacent particles, and communicates motion to them in the same manner, or becomes itself a centre of disturbance.

The principle of Huyghens is the principle of superposition in a generalized form. The disturbance at any point at any instant is due to the superposition of all the disturbances which reach it at that instant from the various parts of the surrounding medium.

Let the points of the surface *mn* be centres from which waves proceed for a short distance *cd*. Then with these centres and a radius *cd*, describe semi-circular waves. The number of such waves being indefinitely large, they will ultimately coalesce to form the new surface *m'n'*, which is the envelope of all the small secondary waves. The effective part of each secondary wave Huyghens supposed confined to that portion which touches the envelope.

The energy of *mn* is thus passed on to *m'n'*, and in the same manner from *m'n'* to *m''n''*, etc.

But the question arises, will not the disturbance be propagated backwards as well as forwards by these secondary wavelets? The answer is that each secondary wave is limited in the same manner as the primary wave, or the agitation of any point, like a pulse on a stretched cord, causes the agitation of points in advance of it, but of none in the direction from which it has come. The law of intensity at each point of a secondary wave has been investigated by Stokes, who has shown that the effect of an elementary wave at any external point varies as $(1 + \cos \theta)$, where θ is the angle between the normal to the primary wave and the line joining the point to the centre of the elementary wave. This quantity vanishes when $\theta = \pi$, or for points directly behind the wave. The disturbance due to a secondary wave, therefore, varies from a maximum at its forward apex to zero at the opposite point in its rear.

135. Reflection of a Plane Wave at a Plane Surface (D., 117). — Let AB (Fig. 78) be a portion of the plane advancing wave, and let CD be the reflecting surface. If AB had met with no obstruction it would have taken the position $A'B'$ at the instant when B arrives at B'. But A becomes a new centre of disturbance which travels backward in the first medium. Then with A as a centre, and with a radius AA', equal to BB', describe a

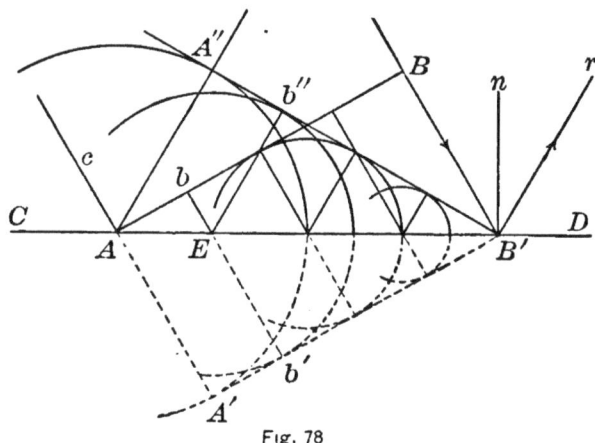

Fig. 78

circle. This circle limits the distance to which the disturbance has spread. In the same time the disturbance from b would have reached b' without obstruction, but it travels first to E, and is then reflected. We must, therefore, draw about E a circle with radius Eb'. In the same manner draw any number of circles.

Finally, from B' draw a tangent to the first circle; it will touch all the other circles, and will be the reflected wave-front. Draw AA'' to the point of tangency with the first circle. Then $AA''B'$ is symmetrical with respect to $AA'B'$; and since $ABB'A'$ is a parallelogram, the triangle $AA'B'$ is equal to ABB'. Therefore, since the triangles $AA''B'$ and ABB' are both equal to $AA'B'$, they are equal to each other, and the angles BAB' and $A''B'A$ are equal to each other. But the former is the angle of incidence,

since the lines BA and $B'A$ are perpendicular respectively
to BB' and nB', and the angle $BB'n$ is the angle of incidence.
In the same way it can be shown that the angle $A''B'A$ is
equal to $nB'r$, the angle of reflection. The angle of incidence, therefore, equals the angle of reflection. The
former is the angle between the incident wave-front and
the reflecting surface CD; the latter is the angle between
the wave-surface after reflection and the reflecting surface. The reflection of sound then follows the ordinary
laws of the reflection of waves.

**136. Relations of the Centres of the Direct and Re-
flected Systems of Waves.** — Let O (Fig. 79) be the
centre of the incident spherical waves, and let them be
reflected from the surface AB. If these waves had met
with no obstruction they would have taken positions at
equal successive time-intervals indicated by the dotted
lines; but they are reflected so as to have the positions of

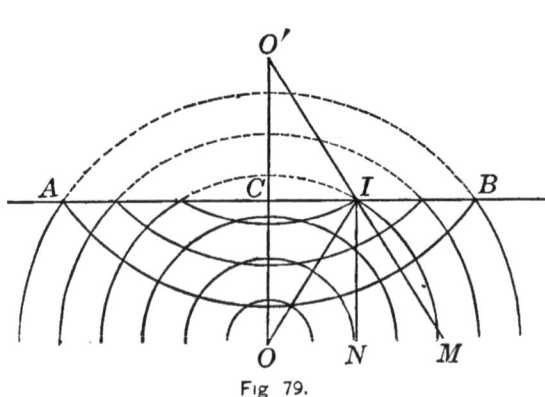

Fig 79.

the full lines sym-
metrically situated
on the other side of
the reflecting sur-
face. Let OI be a
sound-ray, or the
direction of motion
of any point of the
incident wave; draw
IM so that IM and
OI shall make equal
angles with the normal. Then is IM the path of the
reflected ray; that is, it is a normal to the reflected waves.
Project IM backward till it intersects at O' the normal to
the reflecting surface through O. Then is O' the centre

of the reflected waves. The triangles OIC and $O'IC$ are equal. Hence OC and $O'C$ are equal, and the centres of the incident and reflected waves are on a perpendicular to the reflecting surface and equidistant from it. The sound-centre and sound-image are symmetrically situated with respect to the reflecting surface.

137. Echo (D, 425). — Whenever sound passes from one medium to another of different density a part of the energy is transmitted and a part reflected. The one system of waves gives rise to two distinct systems, and the intensity of sound in either direction is weakened by the new medium. In general the energy or intensity of the reflected system increases with the difference in density of the two media. A dry sail, for example, will transmit a part of the sound, and will reflect a part; but if wetted it becomes a better reflector, and almost impervious to sound.

A flame is known to be a fairly good reflector of sound, and the hot air above the flame reflects nearly as well as the flame itself. It is evident from this fact that if the air in clear weather has ascending and descending currents, differing in temperature from the neighboring masses, the sound will be partly reflected and partly transmitted. Its intensity will then fall off with distance much more rapidly than if the air were of uniform density. Sound is often heard more plainly in foggy or rainy weather than when the atmosphere is clear, because then the air is more uniform. For the same reason sounds are heard much further in a quiet night than by day. Humboldt remarks on the great distance that the falls of the Orinoco in South America can often be heard by night; and arctic travellers relate that in the long night of the polar regions a slight sound can be heard an incredible distance.

The most important illustration of the reflection of sound-waves occurs in the case of echoes. Assume that the sensation of sound persists for about one-tenth of a second, during which time sound travels 33 metres.

If then the distance of the reflecting surface exceeds 16.5 metres, the observer may hear the direct and reflected sounds separately. This repetition of sound by reflection is called *echo*. Parallel reflecting surfaces at suitable distances produce multiple echoes. Reflection of sound takes place from buildings, rocks, woods, and hills. If a person can utter five syllables a second, standing opposite a large reflecting surface, at a distance of 165 metres, he can then hear the five syllables repeated by reflection entirely distinct from the original words. For shorter distances the direct and reflected sounds become confused. Such is the case in rooms with bad acoustic properties. A circular building with a hemispherical dome, like the Baptistry at Pisa, may prolong a sound for many seconds by successive reflections. The effect is made more conspicuous by the good reflecting surfaces of polished marble. A single loud sound in the Baptistry at Pisa continues to be audible for twelve or fifteen seconds.

In a similar way the sound of a whistle or a gun on the water is often heard to roll away apparently to a great distance. These are called aërial echoes. A curious echo of this kind was observed by Tyndall during a course of experiments near Dover to determine the effectiveness of sound-signals during a fog. An atmosphere perfectly transparent from an optical point of view may have an acoustic opacity almost impenetrable. If considerable masses of invisible vapor rise from water they may become obstacles for the transmission of sound, by creating heterogeneous layers or banks, at the limiting surfaces of which sound will be partially reflected. A portion of the sound transmitted by one bank is then reflected by the next, giving rise to a curious prolongation of a short signal.

PROBLEMS.

1. Find the wave-length in air of a note due to 128 vibrations per second when the temperature is 20° C.

2. An express train passes a station at a speed of 70 kilometres an hour, and blows a whistle, the frequency of which is 750 vibrations per second. What will be the difference in pitch of the note to an observer at the station as the train approaches and as it recedes, temperature 20° C. ?

3. A stone is dropped into a well and is heard to strike after 3 seconds. Determine the depth, the velocity of sound being 335 metres per second.

4. If the velocity of sound is 332 metres per second, find the number of vibrations which a C fork, with a frequency of 256, will make before the sound is audible at a distance of 50 metres.

5. Three observers are stationed 2, 4, and 6 kilometres respectively from a gun, which is fired at noon. At what time will the report be heard by the several observers if there is no wind and the temperature is 20° C. ?

6. If in the preceding example the wind is blowing at a speed of 80 kilometres an hour, at what time will the report be heard by the first observer stationed directly windward and the second directly leeward ?

CHAPTER VII.

PHYSICAL THEORY OF MUSIC.

138. Musical Intervals. — The *pitch* of a musical sound is the pitch of its gravest component, or fundamental tone; and this depends upon the frequency of the fundamental vibrations of the sounding body.

Pitch may be defined in two ways:

1. *Physically*, as the number of vibrations per second in the lowest tone of the sound.

2. *Musically*, by referring the sound to its place in an arbitrary scale of pitch in use among musicians.

When two notes are sounded together or in quick succession, the ear recognizes a special relationship existing between them, involving a perception of their relative pitch.

This relationship is expressed as a ratio between their frequencies of vibration, and is entirely independent of the absolute pitch of the two tones. It is called a *musical interval*.

Many of these ratios have definite names in musical nomenclature. Thus the ratio of one to one is called *unison;* of $\frac{2}{1}$, an *octave;* of $\frac{3}{1}$, a *twelfth;* of $\frac{3}{2}$, a *fifth;* of $\frac{4}{3}$, a *fourth;* of $\frac{5}{4}$, a *major third;* of $\frac{6}{5}$, a *minor third.* Numerically a musical interval is always equal to or greater than unity. Musical intervals are equal to each other when their constituent notes have the same *relative* vibration-rates.

139. The Diatonic Scale or Gamut. — When three notes, whose vibration-rates are as 4: 5: 6, are sounded together an effect is produced which is pleasing to the ears of Western nations, as distinguished from those of the Orient. Such a combination of three tones is called a *major triad;* and, together with the octave of the lowest of the three, they compose a *major chord.* The perfect diatonic scale is derived from three sets of such triads.

If the three tones have vibration-frequencies as 10: 12: 15, they compose a *minor triad;* and with the octave of the lowest, a *minor chord.*

A single tracing point, like the graver on the diaphragm of a phonograph, may be set in motion by two or more systems of sound-waves simultaneously. If then the surface on which the curve is to be inscribed is moved at right angles to the motion of the tracing point, the resulting curve will be due to the superposition of the several motions in the same plane. The upper curve in Fig. 80

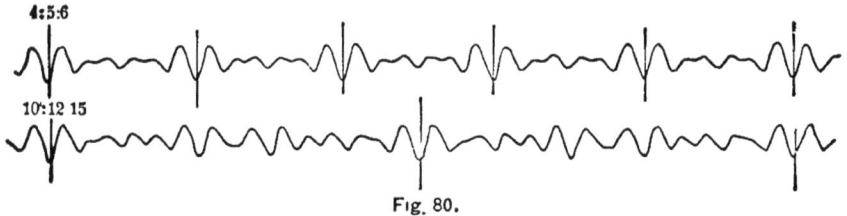

Fig. 80.

is the result of combining in this way three simple harmonic motions constituting a major triad; the lower curve in the same way shows the composite motion resulting from a minor triad. It will be noticed that a complex waveform recurs in each case, but less frequently in the second combination than in the first.

The eight notes of the scale are represented by the letters C, D, E, F, G, A, B, c.

The three major triads are

$$\left.\begin{array}{c} C:\ E:\ G \\ G:\ B:\ d \\ F:\ A:\ c \end{array}\right\}^{1} ::4:5:6.$$

When the several notes of the scale are thus related they give the most pleasing chords.

If then C is due to m (about 64) vibrations per second, the vibration-rates of the other notes of the scale may be found by simple proportion from the above relations.

Thus

$$\frac{E}{C}=\frac{5}{4};\text{ hence } E=\frac{5}{4}\,C=\frac{5}{4}\,m.$$

$$\frac{G}{C}=\frac{6}{4};\text{ hence } G=\frac{3}{2}\,C=\frac{3}{2}\,m.$$

$$\frac{c}{A}=\frac{6}{5};\text{ hence } A=\frac{5}{6}\,c=\frac{5}{6}\,.\,2m=\frac{5}{3}\,m.$$

Pursuing the same method throughout, the following numbers are found to represent the relative vibration-frequencies of the several notes of the gamut:

Vibration No.	64	72	80	$85\frac{1}{3}$	96	$106\frac{2}{3}$	120	128
Name of Note	C	D	E	F	G	A	B	c
Vibration-rate	m	$\frac{9}{8}m$	$\frac{5}{4}m$	$\frac{4}{3}m$	$\frac{3}{2}m$	$\frac{5}{3}m$	$\frac{15}{8}m$	$2m$
Intervals		$\frac{9}{8}$	$\frac{10}{9}$	$\frac{16}{15}$	$\frac{9}{8}$	$\frac{10}{9}$	$\frac{9}{8}$	$\frac{16}{15}$

If the fractions representing the vibration-frequencies are reduced to a common denominator, the numerators may then be taken to denote the relative vibration-frequencies of the eight notes. They are

$$24,\ 27,\ 30,\ 32,\ 36,\ 40,\ 45,\ 48.$$

---

[1] The letters denoting the notes are here made to stand for the vibration-frequencies also.

If the above intervals between the successive notes of the scale are examined, it will be seen that there are only three different ones throughout the perfect diatonic scale. The intervals $\frac{9}{8}$ and $\frac{10}{9}$ are called whole tones, and $\frac{16}{15}$ a half tone, or a *limma.* The difference between the two whole tones is $\frac{81}{80}$, a *comma.* This is of course the ratio between $\frac{9}{8}$ and $\frac{10}{9}$. If, for example, the interval from m to n is $\frac{10}{9}$ and from m to r, $\frac{9}{8}$; then the interval from n to r is $\frac{81}{80}$.

The intervals between C and each of the other notes in succession are called a second, a major third, a fourth, a fifth, a major sixth, a seventh, and an octave. The minor intervals are counted backward from the last note of the scale. Thus the interval between A and c is a minor third.

140. Minor Chords and Transition. — The interpolated notes, additional to the eight of the diatonic scale, are rendered necessary in order to provide for minor chords and to be able to pass from a scale in one key to that in another, a process which is called *transition.* The middle note of a minor triad is lower than that of a major by an interval of $\frac{25}{24}$. Three interpolated notes become necessary in the key of C, viz., three notes below E, B, A, by the above interval.

But if we suppose the gamut to begin with G, then the seven other notes must follow with the same succession of intervals as in the key of C; that is, $\dfrac{9}{8}$, $\dfrac{10}{9}$, $\dfrac{16}{15}$, etc.   Or in other words for the key of G, the three sets of major triads are

$$\left.\begin{array}{l} G:\ B:\ d \\ D:\ F:\ A \\ C:\ E:\ G \end{array}\right\} :: 4:5:6.$$

Comparing these with the three triads for the key of C it will be seen that two of them are identical, while the third contains two notes, F and A, differing from the scale in the key of C.   A numerical comparison of the two scales shows the exact difference.

Key of C.

c', d', e', f', g', a', b', c'', d'', e'', f'', g'',
256, 288, 320, 341⅓, 384, 426⅔, 480, 512, 576, 640, 682⅔, 768

Key of G.

256, 288, 320, 360, 384, 432, 480, 512, 576, 640, 720, 768

The interval between the a's of the two scales is $\dfrac{80}{81}$, while the interval between the f's is much larger.   One other new note besides these two is necessary to provide for minor triads.   But other keys are employed also, some introducing a still larger number of extra notes; so that, with all the naturals as key-notes, the scale would comprise at least 72 notes to the octave.

141. Tempered Scales (D., 390; Bl., 137). — Every transition from one key to another more remote from C multiplies the demand for new tones.   The number of notes required to provide for scales in all keys is far in excess of possible provision in an instrument with fixed keys like the piano. Hence some system of accommodation must be adopted by which the number of notes shall be much reduced by changing the values of the intervals. Such a modification of the notes is called *tempering.* Every system of tempering changes slightly the pitch of each note, so as to bring together into one all the interpolated notes falling between any two adjacent ones of the diatonic scale. The intervals from *E* to *F* and from *B* to *C* being already semitones, no others are interpolated there. The extra notes, therefore, occur in groups of threes and twos, represented by the black keys on the piano, making thirteen notes in the scale, with twelve intervals.

The system of temperament most commonly applied to the organ and piano is known as the system of *equal temperament* introduced by Bach. It makes all intervals from note to note equal, and interpolates only one note in each whole tone of the diatonic scale. Each interval of a half tone equals $\sqrt[12]{2}$ or 1.05946. The result differs widely from pure intonation. On a pianoforte the thirds, three of which are forced to make an octave, are too sharp, though their sharpness adds somewhat to the brilliancy of the music.

The difference between the eight notes in the natural scale of *C* and the equally-tempered scale of *C* appears from the following table :

	C	*D*	*E*	*F*	*G*	*A*	*B*	*c*
Natural . .	1000	1125	1250	1333.33	1500	1666.66	1875	2000
Tempered .	1000	1122.46	1259.92	1334.84	1498.31	1681.79	1887.75	2000

The above numbers represent only relative vibration· frequencies.

" Music founded on the tempered scale must be considered as imperfect music, and far below our musical sensibility and aspirations. That it is endured, and even thought beautiful, only shows that our ears have been systematically falsified from infancy." [1]

It is an incorrect scale, "born of transition in order to avoid the practical difficulties of musical execution."

142. Laws of the Transverse Vibration of Strings (D., 400; A. and B., 375; V., II, 166). — When a disturbance is produced at any point of a stretched string it runs in both directions to the fixed ends from which it is reflected, and passing back on the opposite side is again reflected, and finally arrives at the starting point. The string has then returned to its initial condition of disturbance, it has executed one complete vibration, and each half of the pulse has traversed the length of the string twice. But the wave-length *along the string* is the distance travelled in the period of one complete vibration. If then l is the length of the string

$$\lambda = 2l,$$

or the wave-length for the fundamental tone is twice the length of the string. This wave-length has no relation to the wave-length of sound in air.

Suppose a long, slender, and perfectly flexible string, without elasticity properly speaking, to be strongly stretched and to be drawn aside slightly from its initial position of rest. Then the force tending to restore it to its position of equilibrium is the component of the tension resolved in a direction at right angles to the length of the string. The displacement being small, this force of

---

[1] Blaserna's *Theory of Sound*, p. 140.

restitution varies directly as the displacement, and the motion is simple harmonic.

The coefficient of restitution therefore takes the place of elasticity in the formula for the transmission of longitudinal vibrations (117). We may then write

$$v = \sqrt{\frac{k}{d}},$$

in which k is the tension in dynes per square centimetre of cross-sectional area of the string, and d is the density.

If T is the tension in grammes, then $k = \frac{Tg}{\pi r^2}$, r being the radius of the cylindrical string. The formula for the velocity of the pulse along the string then becomes

$$v = \sqrt{\frac{Tg}{\pi r^2 d}}.$$

We may now put t for the tension in dynes Tg, and m for $\pi r^2 d$, the mass per unit length of the string. Then

$$v = \sqrt{\frac{t}{m}}.$$

But the vibration rate n equals $\frac{v}{\lambda}$, and λ equals $2l$. Hence, substituting,

$$n = \frac{1}{2l} \sqrt{\frac{t}{m}}.$$

The number of vibrations per second of such a stretched string, for its fundamental or gravest tone, is

1. Inversely proportional to its length.

2. Directly proportional to the square root of the tension.

3. Inversely proportional to the square root of its mass per unit length.

These theoretical laws are found to be very exactly

true for long, flexible cords, strongly stretched, and particularly if they are not metallic. But if the cords are short, thick, and lightly stretched, the number of vibrations is always higher than the theoretical number, and it is higher the greater the rigidity of the cord (Violle, II, 188). This rigidity acts in effect somewhat like another tension added to the stretching force T, although the assimilation of the rigidity to a constant tension is not entirely exact.

The mathematical theory for the establishment of the preceding formulæ assumes:

1. That the transverse dimensions of the cord shall be so small that it can be regarded as a simple, absolutely flexible thread.

2. That the cord is sufficiently stretched and is only so slightly deformed that the variable forces of elasticity resulting from these deformations may be completely neglected relative to the permanent tension T.

143. The Vibration of Strings in Segments (T., 93; V., II, 169; Z., 154). — A string is capable of vibrating not only as a whole, but also in equal segments; and the number of such segments, when it is made to vibrate in a single mode, depends upon the relation between the periodic time of the disturbances applied to it and the speed with which these disturbances travel along the string.

Take a soft, thick, twisted cotton cord, from five to ten metres long, and fix one end to a firm support. Holding the free end in the hand, move the hand gently up and down to find the natural period of oscillation of the cord as a whole, or in one segment. When this is found a series of slight impulses, so timed as to aid the oscillations of the cord, will cause it to swing through a wide amplitude.

Next, apply the transverse impulses twice as often, keeping the cord stretched with the same tension. It will now divide into two equal segments. If the motion of the hand is in a circle, the two segments of the cord will be large spindles with almost no motion at the middle point. Then let the impulses be three times as fast as at first ; the cord will divide into three vibrating segments and will have the appearance of Fig. 81. The two ends and the

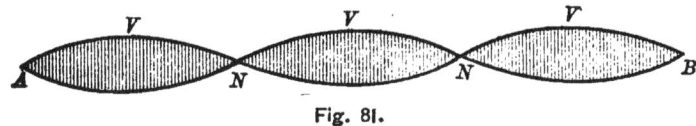

Fig. 81.

points, *N, N,* are called *nodes.* They are the points of least motion. The intermediate points, *V, V, V,* are called *antinodes.* Two points on opposite sides of a node are always moving in opposite directions. If the motion of every point of the string is circular then, while two points on opposite sides of a node are moving in the same dircetion around their respective circles, one is moving in one direction in space on one side of a circle while the other is moving in the other direction on the other side of its circle. In other words, their motions differ in phase by half a period.

By increasing the frequency of the movements of the hand the cord may be made to divide into four, five, six, or even more segments, according to the dexterity of the experimenter.

When a cord vibrates in this way, with fixed nodes, it illustrates what is known as *stationary waves.* Stationary waves result from the superposition of two wave systems, one direct and the other reflected. The relation between the speed of transmission of the wave along the cord and the number of vibrations is easily found.

Let a transverse disturbance be started at one end; it runs along the cord and is reflected at the other end with a change of sign of the motion, a protuberance being transformed into a depression. On arriving again at the origin or free end it is again reflected with a change of sign. If now this pulse, which has been twice reflected, agrees in phase with another pulse just starting from the origin, then their motions will be added together; and in this way a periodic movement applied at one end is rapidly amplified, the wave twice reflected being identified with the direct wave, if the period of the double passage of the wave along the cord is a whole number p of periods of vibration. Let v be the speed of transmission of the transverse motion along the cord, l the length of the cord between the two points of reflection at the ends, and T the period of vibration. Then the condition of reënforcement is

$$\frac{2l}{v} = p\,T.$$

But since n, the vibration number, is the reciprocal of T,

$$n = p\frac{v}{2l}.$$

If p is unity, the cord is vibrating in a single segment; if p is two, the cord divides into two segments and the period is half as great as before. If p is three the cord divides into three segments, and each segment executes three complete vibrations while the pulse travels over twice the length of the cord.

144. Segmental Vibration of the Monochord. — A monochord is essentially a single stretched wire. It is usually mounted on a box of thin resonant wood, with lateral apertures communicating with the external air. Near the ends of the box are bridges and the wire is stretched over them. A sonometer, as the instrument is often called, provided with three wires, is shown in Fig. 82.

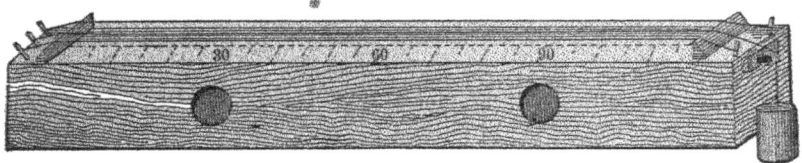

Fig. 82.

For certain purposes it is better to take a steel piano wire, about No. 22 gauge, and four metres long, and stretch it over two appropriate bridges attached to the top of a long table. With this wire properly stretched, the vibration in segments may be strikingly illustrated. A thin piece of cork, about an inch in diameter with a small hole at the centre, should slide readily along the wire. Provide little riders of stiff paper bent double, some white and some red. Let the slip of cork be placed one metre from one end, and let white riders be placed at the middle and at one metre from the other end, while red ones are mounted on the wire at intermediate points. Now touch the cork very lightly and draw a heavy bow across the short division of the wire — the metre length at one end. If this is deftly done, the wire will sound, the red riders will be violently unhorsed, while the white ones will remain in place on the wire. The white riders therefore mark the place of the nodes, and the string vibrates in four segments, each a metre in length. If the cork slip is placed at 80 cms. from one end and the white riders are mounted at distances of 80 cms. apart along the wire so as to mark the places of five equal divisions, then, upon agitating the first segment of the wire by the bow as before, the intermediate red riders will be thrown off, while the white ones will remain sitting. This method of exhibiting nodes and antinodes was first employed apparently by Noble and Pigott at Oxford in 1673, but the application of it to a monochord was made by Sauveur in 1701.

In this experiment impulses of the proper period are obtained by the vibration of the short segment of the wire, which must of course then be an aliquot portion of the whole. This segment furnishes the timed impulses for the remainder of the wire, and the position of the cork slip must be considered the origin, corresponding with the free end of the cotton cord held in the hand as already described. The relationship between period and speed of transmission obtained in the last section must then clearly hold true for this case of the monochord.

The origin, marked by the cork, which serves to produce a node at the point, is not a place of no motion, but of minimum motion; the small movements transmitted across it are accumulated and amplified in the other segments by the addition of the direct waves to the reflected ones.

A stretched wire or string may thus vibrate in any number of equal segments. The number of vibrations executed per second will be proportional to the number of segments into which the wire divides. Thus the vibration-frequency for three segments will be three times as great as for the fundamental of one segment; for four segments, four times as great, etc.

145. Melde's Experiments (V., II, 170; D., 405; Z., 157). — A long white silk cord is stretched horizontally between a small fixed pulley and a vertical tuning-fork. The plane of the two branches of the diapason contains the cord, the end of which is displaced longitudinally when the fork vibrates. The cord relaxes, falls to its lowest position with the forward movement of the fork, again rises to the horizontal, and then to its highest position, when the fork is again in its most forward position. The longitudinal movement of the point of attachment thus gives rise to a transverse motion of the cord, with a period double that of the fork. The entire cord will oscillate then an octave below the tuning-fork. For this purpose the tension must be carefully adjusted by weights in a scale pan hung on the cord beyond the pulley. When the exact tension required has been found the cord spreads out in a pearl-white spindle, which appears to be perfectly fixed and stable.

If now the fork be turned on its axis so that it communicates transverse impulses to the cord, the conditions then obviously require the fork and cord to vibrate in unison. The cord will then break up into two segments separated by a node. Each half vibrates twice as fast as the entire cord, and so keeps in unison with the fork. This demonstrates the law of lengths.

Next turn the fork back into its former position. By reducing the weights, including the pan, to one-quarter, the cord again divides into two segments. Each segment again vibrates at a rate equal to an octave below the fork. As a whole string it would vibrate two octaves below; but by dividing into two segments its rate remains the same as at first, while the time required for the pulse to travel twice its length has been doubled by reducing the tension.

With the tension reduced to one-ninth the cord divides into three segments. Since in each case the segments vibrate an octave lower than the fork, and since for a tension of one-ninth, for example, the cord divides into three segments, or the vibration-frequency is reduced to one-third, the law of tensions is thus verified.

PROBLEMS.

1. Suppose a string, vibrated by a tuning-fork, is stretched with a weight of 270 gms. and divides into four segments. What must be the weight to cause it to divide into three segments with the same fork?

2. A cord attached to a fork with its plane of vibration in the direction of the string divides into two segments when stretched with 270 gms. With the plane of vibration of the fork at right angles to the cord, what weight must be applied to cause it to divide into three segments?

3. A cord vibrates synchronously with the attached fork by dividing into three segments. If it be replaced by a similar one of the same length and four times the sectional area, what relative weight will be required to cause it to divide into four segments?

146. Overtones. — When a stretched string or wire is made to vibrate it not only gives its fundamental tone, but it divides at the same time into one or more sets of equal segments, which produce higher tones than the fundamental; so that usually there are several series of vibrations superposed upon the fundamental one. The tones of higher pitch associated with the fundamental are known by the general name of *overtones* or *upper partials*.

In the case of strings the division will be into two, three, four, five, etc., equal segments, with vibration-frequencies two, three, four, five, etc., times that of the fundamental tone. The intervals between the fundamental and the overtones are therefore an octave, an octave plus a fifth, or a twelfth, a double octave, two octaves plus a major third, two octaves plus a fifth, etc. If, for example, the fundamental is *C*, the first overtone is *c'*, the second *g'*, the third *c''*, the fourth *e''*, and the fifth *g''*, while the sixth overtone having a frequency of vibration seven times the fundamental, is not represented by any tone in the diatonic scale or gamut. The eighth overtone, with nine times the frequency of the fundamental, is *d'''*, which produces a discord with the fundamental tone. The formation of these particular overtones is prevented on the pianoforte by having the hammer strike the wire at a distance of a little less than one-seventh the length of the string from one end. Since the point struck must be an antinode for every system of subdivision, all modes of segmental vibration requiring a node at the point struck are thereby eliminated.

147. Distinction between Partial Tones and Harmonics.[1] — When the vibrating parts of a musical instrument which is producing composite tones, such as the strings of a piano or a violin, or the column of air in an organ pipe, divide into several series of segments at the same instant, all the tones produced by these segmental vibrations are called *partial tones*. But the vibrating parts of many musical instruments may execute a variety of very complex or imperfectly pendular motions, which are not made up by the superposition of several series of equal subdivisions. All such vibrations, however, in order to produce musical sounds, must have the characteristic of periodicity; that is, they must repeat themselves over and over in certain definite and equal intervals of time. Such complex periodic motions are subject to the following law of Fourier : *Every periodic motion whatsoever may always be considered as the resultant of the superposition of a*

---

[1] Koenig's *Quelques Expériences d'Acoustique,* 218; Wiedemann's *Annalen,* **1881.**

definite number of pendular vibratory motions, or is always resolvable into a definite number of commensurate simple harmonic motions. The frequencies of these simple harmonic components of the complex periodic motion are all *exact multiples* of the fundamental; that is, if the period of the fundamental tone is T, then the periods of the overtones must all fall in the series $\dfrac{T}{2}$, $\dfrac{T}{3}$, $\dfrac{T}{4}$, $\dfrac{T}{5}$, etc., or their vibration-frequencies are 2, 3, 4, 5, etc., times that of the gravest tone. Now the component tones due to these higher frequencies, which are rigorously exact multiples of the fundamental, Koenig calls *harmonics.* It is true that the word harmonic is often applied to those partial tones which harmonize with the fundamental, but that is not the meaning attached to the term here. Koenig says: " Among the sounds into which the sonorous mass, which emanates from a vibrating body, may be resolved, we may distinguish harmonics and partial tones. These last have their origin when the body in question executes simultaneously several modes of vibration which it can adopt separately, as in the case of the string; while the harmonics are due to the resolution into simple pendular motions of the imperfectly pendular oscillations of the sounding body executing a single mode of vibration." Partial tones are therefore due to the actual subdivision of the sonorous body into vibrating segments; harmonics, on the contrary, are the commensurate components of the motion when the body vibrates periodically, but by only one mode, and that a complex one. Harmonics have frequencies which are necessarily exact multiples of the fundamental; the frequencies of partial tones may or may not be exact multiples. Even if they are exact multiples, still they originate, each in a corresponding subdivision of the sonorous body or source of sound, while the harmonics have no such physical foundation. They are the components into which mathematical analysis shows that the imperfectly pendular motion is resolvable.

Harmonics are always due to frequencies represented by the series of exact whole numbers, while the frequencies of partial tones approach only more or less nearly to their theoretical values. Two diapasons, whose fundamentals are very exactly in unison, may give partial tones of the same order which produce loud beats, and which are therefore not in unison. If the overtone of one of the diapasons is an exact multiple of its fundamental, that of the other cannot be.

Partial tones then are not rigorously exact multiples of the fundamental in respect to their frequencies of vibration. The frequencies of vibration of all the segments into which a string, for example, divides are not necessarily or exactly equal to each other. The variation from such equality may be due to variation in cross-section of the wire or string, to variation in density or hardness, in the physical or chemical state of the carbon present in different parts of the wire, etc. The wire presenting the greatest uniformity in all respects throughout its length will give the best tone by producing partial tones which are as nearly as possible multiples of the fundamental.

148. The Transverse Vibration of Rods (V., II, 195; B., 235; D., 407). — Rods vibrating transversely may have a circular, square, or rectangular section. A rod of circular section vibrates transversely in all directions without difference. One with a rectangular section vibrates with larger amplitude in a plane at right angles to the broad face than in the plane parallel to this face.

In the transverse vibration of rods, unlike that of strings, the force of restitution is the elasticity of flexure. The theory is complex, but the number of vibrations per second is given by the equation

$$n = C \frac{t}{l^2} \sqrt{\frac{e}{d}}.$$

C is a constant which depends upon the manner of supporting the rod. If the rod is free or clamped at both ends C is 1.78; if free at one end only it is 0.28. For the other terms, t is the thickness of the rod, l its length, e its coefficient or modulus of elasticity, and d its density.

For rods of the same thickness the frequency of vibration is inversely as the square of the length; but if the thickness and the length vary in the same ratio the frequency is inversely as the length. A tuning-fork five cms. long gives a note an octave above one ten cms. long, provided the two forks have the *same relative dimensions.* The same rule applies to reeds.

The vibration-frequency of a rod is independent of its width, but is directly proportional to its thickness. Hence if two rods or bars of the same material have the same length, while one is twice as thick as the other, the thick one will vibrate in half the period or with twice the frequency, whatever may be their relative widths.

The partial tones of rods rise much more rapidly than those of strings. For a rod fixed at one end and free at the other Chladni found the following relative frequencies:

	1	6.25	17.5	34.25	56.5	84
or	$(1.2)^2$	3^2	5^2	7^2	9^2	11^2

Examples of the use of transversely vibrating rods in musical instruments are the reeds of an accordion or har monium, the tongue of a jew's-harp, or of a music box, the reeds of reed-pipes in organs, the claque-bois or xylophone, and the tuning-fork.

The xylophone is a primitive instrument with rods free at both ends. It consists of a series of small wood prisms of convenient length and thickness, supported by strings at the nodes, which are about one-quarter of the length from each end. The prisms are adjusted to give the notes of the scale. They are played by striking them in the middle with a light hammer having a soft elastic face.

149. The Diapason or Tuning-Fork. — The tuning-fork is one of the most important applications of vibrating rods free at both ends. A straight elastic bar when sounding its lowest note has two nodes each at a distance from the end of about one-fourth the distance between them. As this rod is gradually bent into the form of a tuning-fork (Fig. 83), the nodes approach each other; and when the fork is provided with a

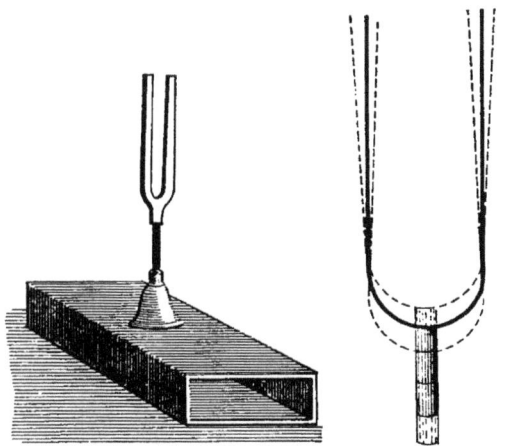

Fig. 83.

stem the nodes are near the bottom of each branch. The two branches then vibrate in unison, each comporting itself sensibly like a rod free at one end and fixed at the other. The stem or base of the fork has a slight up and down motion, which is transmitted to the resonant box on which it is mounted.

The vibration-frequency of the fork is independent of the breadth of the branches, but is directly proportional to their thickness measured in the plane of vibration, and inversely proportional to the square of their length,

or
$$n = K \frac{t}{l^2},$$

where K is a constant which equals for steel about 82,000, the unit of length being the centimetre.

The partial tones succeed one another according to the law of Chladni already given, viz.,

1, 6.25, 17.5, 34.25, 56.5.

The first partial tone of a fork, giving for its fundamental 256 vibrations per second, will be $256 \times 6.25 = 1600$. The first overtone corresponds to the presence of two nodes on each branch, the second overtone to three, etc. The ratio of the first overtone to the fundamental varies somewhat on different diapasons. Tyndall found these values to be comprised between 5.8 and 6.6.

The diapason is the true standard of the musical scale. That the pitch of the sound of a diapason may be absolutely definite, it is necessary that the amplitude of the oscillations be very small, not exceeding $\frac{1}{100}$ of the length of the branches, and the temperature must remain constant. Koenig adjusts all his diapasons at 20° C. According to Koenig the frequency of a steel diapason diminishes $\frac{1}{8943}$ when the temperature rises 1° C.

The isochronism of the small oscillations of a fork, at a constant temperature, furnishes a very exact chronograph for recording small intervals of time. For this purpose the vibration must be maintained with the same amplitude as nearly as possible.

This is effected readily by means of electricity and an electro-magnet.

150. **The Transverse Vibration of Plates (K., 32; V., II, 229; B., 237; Tyn., 139).** — If a square or a round plate of elastic material, such as glass or brass, be clamped at the centre in a horizontal position, and sand be scattered upon it, this sand will gather along certain definite nodal lines (Fig. 84) when the plate is made to emit sound by bowing on the edge. These sound-figures were first obtained by Chladni, and are known as Chladni's figures. The explanation of them

Fig. 84.

involves many difficulties. For certain of the simpler figures, the explanation given by Wheatstone will suffice here.

Consider a long narrow plate, free at both ends, and vibrating transversely so as to give its fundamental tone. It has then two nodal lines running across it at a distance of 0.224 of its length from the ends. The width of this plate, which is essentially a rod or bar, does not affect its frequency within wide limits. Let us therefore assume a width equal to the length. We have then a square plate with a system of nodal lines parallel to two opposite sides. But unless the plate is clamped entirely across its surface, a disturbance at any point is as likely to start vibrations with nodal lines parallel to one pair of sides as to the other; we may therefore suppose that there are two systems of vibrations superposed on the same plate crossing

each other at a right angle. We have then a new mode of
vibratory motion characterized by a system of nodal lines
passing through the nodal points common to the two
systems, and through those points where the motion is zero
because of the reciprocal action of the two superposed
motions. This method offers certain advantages, in spite
of its defects, for a first or approximate explanation of the
phenomena.

Let the two pairs of dotted lines in Fig. 85 represent the
nodal lines for the two systems of superposed vibrations.
These may coexist in two ways. If the two systems are
superposed as shown in Fig. 85, where the sign + means a
motion upward and − a
motion downward, then
the nodal lines of the
resulting system will be
those of 3, Fig. 86. For
the four points of inter-
section of the two pairs
of rectangular nodal

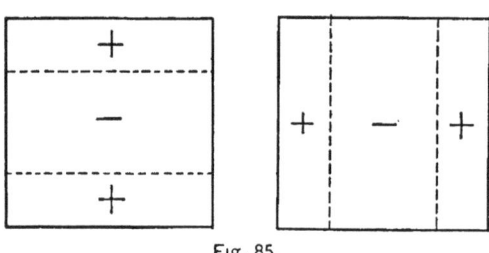

Fig. 85.

lines will be points on the resulting nodes; the middle
point of each side will also be on the new nodal lines,
because the motions of the two systems are then in opposite directions. Con-
necting these points to-
gether the result is the
square of 3, Fig. 86.

If, however, the mo-
tions of one of the su-
perposed systems of
Fig. 85 changes sign,

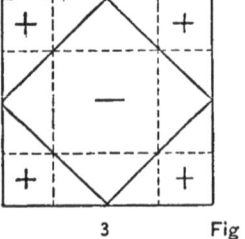
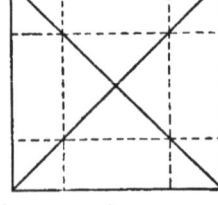

3   Fig 86.   4

then the resultant lines of minimum motion will be the
two diagonals of 4, Fig. 86, since then the corners of the

square will be affected by motions of opposite sign in the two component systems. The former figure, with the angles rounded off, has been obtained by clamping the plate near the middle of one edge, and bowing it at one of the angles.

The other figure is readily produced by clamping the plate at the centre, holding the finger against one corner, and bowing slowly at the middle point of an adjacent edge.

The two systems of superposed waves at right angles may be shown by clamping a glass plate at·the middle, and after carefully levelling, spreading over it a thin layer of water. When the plate is vibrated the surface of the water is agitated with stationary waves in the form of a square check, showing plainly the coexistence of the two rectangular systems of motion. Even a film of soapy water in a square opening will show similar stationary figures when thrown into vibration by any appropriate sound. A triangular opening, covered by a film, gives rise to three sets of plain waves, which, together with their reflected systems, produce stationary waves of a hexagonal pattern.

When a round plate is clamped at its centre, its fundamental tone is produced by a division into four equal segments by two diameters; the first overtone is due to a division into six segments by three diameters, the second by eight segments and four diameters, etc. Adjacent segments, like those of strings, are always in opposite phases of motion.

With rectangular plates whose sides have such relative dimensions that the two wave systems have frequencies represented by some simple ratios, Wheatstone and Koenig have obtained figures approximating very closely to the theoretical Lissajous curves.

151. Resonance (V., II, 279; Z., 266; Bl., 51). — When vibrations come to an elastic body in accord with those which it can itself execute, it is set vibrating as a whole; and, under the repeated action of the synchronous impulses, it may oscillate in complete unison with the external vibrations. This is *resonance*. Resonance depends upon the cumulative effect of small disturbances when applied to a body in such a way as to synchronize with its own motions. One string thus takes up the vibrations of another which has the same vibration-rate. When two heavy pendulums are hung on the same stand and adjusted to swing in exactly the same period, the motion of one of them will be communicated to the other. One drags slightly behind the other and absorbs its energy till the first one comes nearly to rest. The process is then reversed. An organ will often throw windows into loud vibration, producing a rattle. If two tuning-forks, mounted on reënforcing cases, are adjusted to exact unison, the phenomenon of resonance is easily demonstrated at a distance of several metres between them. When one has been bowed and is then stopped by touching it, the other will be found to be producing a very audible sound. The impulses setting the second fork in motion may even be transmitted to a distance by electricity along a wire instead of through the air.

Every elastic body has its own rate of vibration, depending upon its coefficient of elasticity, its density, and its dimensions. A mass of confined or enclosed air has its own period of vibration. Hold a common *A* tuning-fork over the mouth of a tall cylindrical jar, and while the fork vibrates pour in water slowly. As the air column shortens the sound increases in loudness up to a definite point, beyond which the further shortening of the column of **air** in the jar diminishes the sound. If forks of different pitch are tried, each one will be found to have its own length of air column which will reënforce its sound. This increase in the volume of sound, due to the synchronous vibration of another body, usually a mass of partly enclosed air, is resonance.

The resonators of von Helmholtz (Fig. 87) are very valuable for researches in sound. They consist of spheres, provided with two opposite tubular openings; one is short and straight, making free communication with the outer air;

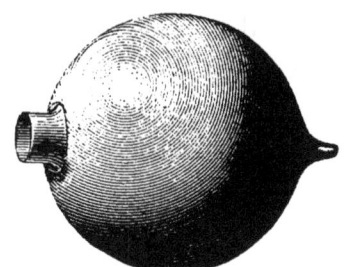

Fig. 87.

the other is small and bell-shaped, so as to be introduced into the ear, where it is closed by the membrane of the tympanum. Each resonator is adjusted in dimensions to respond as a fundamental to some particular tone; it is then practically responsive to only a single tone, and whenever this one is present it declares itself in the ear with a prodigious force. By this means one can distinguish the presence of a feeble sound in the midst of many other loud ones. The "sound of the sea," heard in a sea shell, is a similar reënforcement by selective absorption of vibrations.

152. The Length of Organ Pipes and the Wave-Length of their Fundamental Tone. — If we take several glass or metal tubes of different length and about two cms. in diameter and blow sharply across the edge at one end, while the other end is closed, we shall find that the different tubes give sounds of different pitch, the longer the tube the graver the sound of its fundamental tone. The air in each tube has a definite rate of vibration, and when by blowing across the tube a flutter is produced at one end, consisting of disturbances of various frequencies mingled together, the air column of the tube selects for reënforcement the disturbance of its own rate and exalts that into a musical sound. This reënforcement is accomplished by means of resonance.

For the purpose of analyzing the actions going on in the pipe, suppose a reed vibrating at the middle of a tube closed at both ends (Fig. 88), and that its rate corresponds

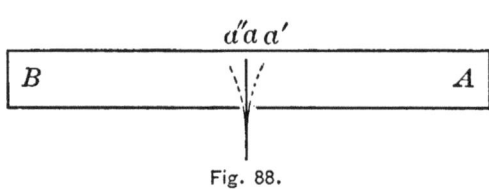

Fig. 88.

with that of the enclosed mass of air, or that it covibrates with it. Let the reed be drawn aside to the position a''. Then while it moves from a'' to a' let the condensation in front of it run from the reed to the end A, and after reflection back to the reed again. Similarly while the reed moves from a' to a'' let the condensation run from it to B and back again to the middle of the pipe. The reed has now executed one complete vibration, the condensation has run over the length of the pipe twice, and the two are ready to repeat the process. The condensation which has been twice reflected is at the middle of the pipe and moving toward A. It is thus ready to join the second condensation produced by the reed and running toward A. The motion of the rarefaction may be similarly traced, the condensation running in one direction while the rarefaction runs in the other. Since the disturbance traverses the pipe twice during a complete vibration of the reed, the wave-length of the sound is twice the length of this pipe.

If now the pipe be cut in two at the middle and the

reed vibrate at the open end, the condensation will run into the pipe and back to the reed during the forward excursion from a'' to a', and the rarefaction will then run in and back during the return movement of the reed from a' to a''. The initial conditions then recur. The disturbance traverses the pipe four times during a vibration of the reed. Such a tube corresponds with a closed organ pipe, which is closed at one end only. The stopped organ pipe is therefore one-fourth the wave-length of its fundamental tone in air.

If the reed be supposed to vibrate at one end of a pipe open at both ends (Fig. 89), then while the reed moves from a'' to a' the condensation runs the entire length of the pipe to A and is there reflected as a rarefaction; that is,

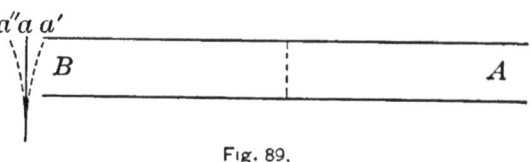

Fig. 89.

the condensation changes sign while the motion of the air particles, by which the rarefaction is propagated backward, has not changed sign. It continues in the direction from B toward A. While now the rarefaction as a reflected wave runs from A toward B, the reed by its motion from a' to a sends another rarefaction into the tube at the end B. The two meet at the middle of the tube, producing a node, and they then pass on to the two open ends of the pipe; so that when the reed reaches a'' the rarefaction has returned to that point to be reflected with a change of sign as a condensation. The reed then sends in another condensation, and the two condensations are concordant. These running into the pipe meet the one reflected from the distant end at the node in the middle.

The disturbance then will be found to traverse the pipe

twice while the reed executes a complete double vibration, or the length of the pipe is half the wave-length of its fundamental sound.

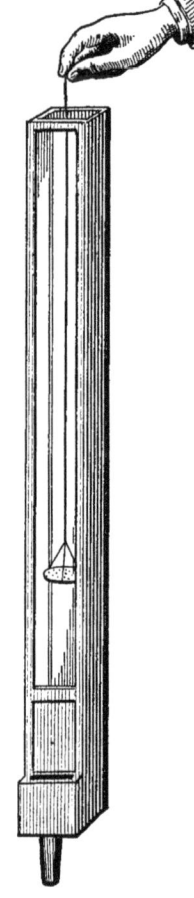

This constitutes an open organ pipe. It has a node near its middle for its gravest tone. If a stopped pipe and an open pipe give notes of the same pitch, the open pipe is twice the length of the closed one.

A node is a place of minimum motion and maximum change of density; an antinode, on the other hand, is a place of maximum motion and minimum change of density. The node at the middle of an open pipe for its fundamental tone may be shown by means of a thin stretched membrane on which some fine sand is strewn (Fig. 90). When this is lowered into the pipe by means of a thread it will buzz, except near the middle where the sand ceases to be agitated.

Fig. 90.

153. Relation of the Overtones to the Fundamental in Open Pipes (V., II, 121). — Since reflection from the open end of a pipe changes the sign of the condensation but not of the motion, a wave twice reflected will have the primitive sign, and will accord with a direct wave if the course traversed by it, or the double length of the pipe $2l$, contains a whole number p times the length of the wave λ. The condition for reënforcement is then

$$2l = p\lambda \, ;$$

or since
$$\lambda = VT = \frac{V}{n},$$

$$n = p\,\frac{V}{2l}.$$

This formula contains all the laws relating to open organ pipes, which, together with those relating to stopped pipes, are known as the laws of Bernoulli. They were established by Daniell Bernoulli in 1762.

When $p = 1$ the formula becomes
$$\lambda = 2l,$$
or the length of the pipe is half the wave-length of the fundamental sound in air.

When p is made successively 2, 3, 4, etc.,
$$\lambda = l,\ \lambda = \frac{2}{3}\,l,\ \lambda = \frac{1}{2}\,l,\ \text{etc.,}$$

or the wave-lengths are represented by the series $1, \frac{1}{2}, \frac{1}{3}, \frac{1}{4}, \frac{1}{5}$, etc., while the vibration-frequencies are proportional to the series 1, 2, 3, 4, 5, etc., both including the funda-

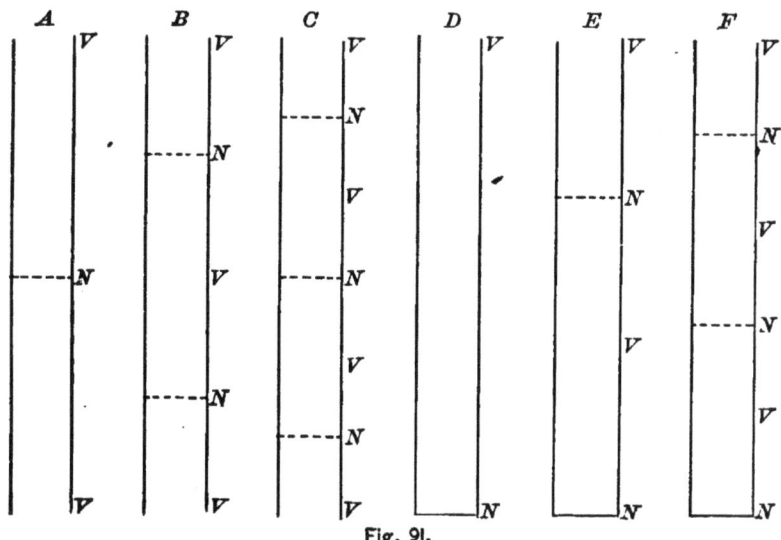

Fig. 91.

mental. The overtones are due to a division of the pipe into vibrating segments as shown on the left of Fig. 91. A is the fundamental with an antinode V at each end and a node N at the middle.

The first overtone adds another node and another antinode, since nodes and antinodes must alternate and succeed each other at equal distances apart. The half vibrating segment in B from either open end to the adjacent node is one-quarter of the length of the pipe and half as long as for the fundamental tone. Its frequency is therefore twice as great.

For the second overtone, as shown in C, still another node and antinode are added, the half segment is now reduced to one-third its primitive length, and the frequency is three times that of the fundamental. The next overtone would require four nodes, the next five, and so on.

154. Relation of Overtones to the Fundamental in Stopped Pipes (V., II, 123). — In the stopped pipe the reflection at the closed end changes the sign of the motion, while the reflection at the mouthpiece changes the sign of the condensation. A wave twice reflected will have the sign both of its motion and its condensation reversed, and it will accord with a new direct wave from the origin if the course traversed, $2l$, *increased by* $\dfrac{\lambda}{2}$, is equal to $p\lambda$. Whence

$$2l = p\lambda - \frac{\lambda}{2},$$

or

$$4l = (2p - 1)\lambda.$$

Therefore

$$n = (2p - 1)\frac{V}{4l}.$$

When $p = 1$, $\lambda = 4l$, or four times the length of the pipe.

When p has successive values 2, 3, 4, etc., $\lambda = \dfrac{4}{3}\ l$, $\lambda =$ $\dfrac{4}{5}\ l$, $\lambda = \dfrac{4}{7}\ l$, etc., in succession, or the wave-lengths are represented by the series $1, \dfrac{1}{3}, \dfrac{1}{5}, \dfrac{1}{7}$, etc.,

and the vibration-frequencies by the series
$$1,\ 3,\ 5,\ 7,\ \text{etc.,}$$
both including the fundamental.

The overtones of stopped pipes are due to a division into segments as represented in *D*, *E*, *F*, Fig. 91.

The first overtone adds one node and one antinode, so that the half segment is one-third as long as for the fundamental, as shown in *E*, and the frequency is three times as great.

For the second overtone, two nodes and two antinodes additional to those of the fundamental are required. The whole pipe is therefore divided into five half vibrating segments, each is one-fifth as long as for the fundamental, and the frequency is five times as great; and so on.

In the successive internodal spaces the motions are always in opposite directions, or of opposite sign. After a half period of the sound produced, these motions are all again equal, but have changed sign. The motions are always of opposite sign on the two sides of a node. Also at any instant successive nodes are affected by alternative condensations and rarefactions; and these all change their signs, or exchange places, every half-period $\dfrac{T}{2}$. At intermediate instants the air is at atmospheric pressure.

155. Experimental Verification. — The experimental verification of the order of overtones in both open and stopped pipes is readily made by means of a series of diapasons having relative vibration-rates of 1, 2, 3, 4, 5, etc.

One must also be provided with two long narrow pipes, whose fundamentals are in unison with the lowest diapason, one open and the other stopped. On such long pipes it is difficult to obtain the fundamental, but one may be assured that it is nearly in unison with the diapason by breathing into it and sounding the fork at the same time. Faint beats may then be perceived if there is a slight difference in the pitch. If the pipe is the open one, then by blowing slightly harder and sounding the second fork an octave higher than the first, audible beats will again be produced, showing that the first overtone has twice the frequency of the fundamental. The second overtone will be found to be in unison or to beat slowly with the third fork, the next overtone with the fourth fork, and so on. The partial tones of the open pipe are thus the octave, the octave plus the fifth, the double octave, etc., of the fundamental.

Similar experiments made with the stopped pipe will show that its first overtone will beat slowly with the third fork of the series, its second with the fifth, etc. The partial tones produced by it are therefore an octave plus a fifth, two octaves plus a third, etc., above the fundamental.

The position of the antinodes may be found by the simple device of piercing the side of the pipe with small holes at the points where the antinodes are for any particular overtone selected. The pressure at the antinode is always atmospheric. An opening made there will not then affect the sound, while the pitch will change if an opening be made at any other point. These small openings in the side of a narrow wood pipe can be covered by turning a small button. Suppose the second overtone of an open pipe is blown. The division into segments is shown in *C* (Fig. 91), and there is an antinode at one-third the length of the pipe from either end. If, then, either button be turned so as to open the pipe at *V*, no change in pitch will be produced for this overtone. For the first overtone the hole may be at the middle without affecting the sound.

With the closed pipe, on the other hand, the opening for the first overtone must be made at one-third the length of the pipe from the closed end, and for the second overtone one-fifth or three-fifths from the closed end, *E* and *F* (Fig. 91).

This experiment demonstrates that there is no change in pressure at the antinodes. Change of pressure occurs, however at all other points, and especially at the nodes. Koenig's "manometric flames" are admirable for illustrating this phase of the phenomena of organ pipes. At the proper

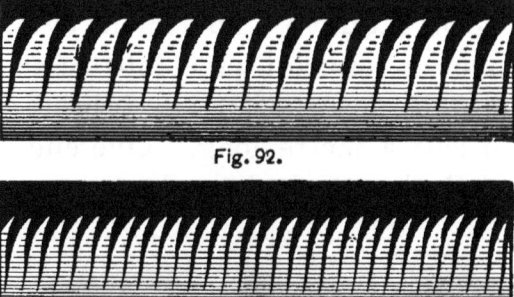

Fig. 92.

Fig. 93.

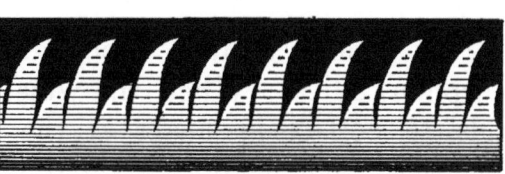

Fig. 94.

points in the side of a pipe holes about three cms. in diameter are covered with a thin diaphragm of gold-

beater's skin or rubber.   Over this is fastened a small
chamber or capsule, into which illuminating gas is admitted.   A small burner is attached, and the flame
is examined by means of reflection from a rotating
mirror.   The membrane takes the motion of the air in the
pipe and communicates it to the gas on the other side
of it.   This change of pressure causes the gas flame to
vibrate in unison with the changes of pressure, and its
image in the rotating mirror is a serrated band (Fig. 92),
which represents the fundamental tone.   If the pressure is
increased so as to produce the first overtone in an open
pipe, there are twice as many tongues of flame as before
(Fig. 93).   By properly regulating the pressure both the
fundamental and the first overtone may be produced simultaneously.   The appearance of the flame in the mirror
is then as represented in Fig. 94.

156. Kundt's Experiment (V., II, 159). — The division of a resonant pipe into segments is most beautifully
shown by means of a glass tube about two cms. in diameter
and half a metre long.   One end is closed and a common
whistle is attached to the other (Fig. 95).   Within the

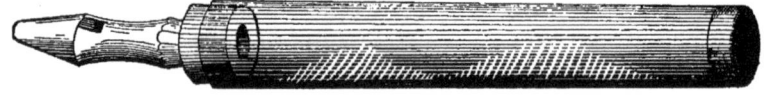

Fig. 95.

tube is placed a little lycopodium powder, or better,
amorphous silica.   When the whistle is blown the powder
tends to collect in heaps at the antinodes, and at the
same time each heap is divided into thin, airy segments
by vertical stratifications.   The agitation is sufficient to
support the powder in opposition to gravity.   The distribution of the powder exhibits the stationary waves due to

the superposed direct and reflected systems, and the stratification shows the shifting of the nodes resulting from a lack of covibration between the whistle and the aerial segments. The subdivision of the pipe changes when the pitch of the whistle changes with increase of pressure.

Kundt has given to this experiment a very elegant form, designed to compare the velocity of sound in air and other media. His apparatus consists of a long glass tube, closed at one end by a cork furnished with a stem, which permits of slight adjustment by forcing in or withdrawing. Into the other end passes a rod securely clamped at its middle and terminated in the interior of the tube by a light disk of a diameter slightly less than the tube. The interior of the tube is lightly powdered with the amorphous silica or fine-sifted cork filings. When the rod is thrown into longitudinal vibration by friction it vibrates precisely like an organ pipe open at both ends, and giving its fundamental tone. The disk on the inner end communicates its displacements to the air enclosed in the tube, and the gaseous column tends to divide into segments of such length that they will all vibrate in unison with the rod. The adjustment for unison is made by moving the stopper till the powder gathers into small detached heaps, which indicate very neatly the exactness of the adjustment. The column of air is then an exact multiple of half wave-lengths, and the distance l between two adjacent nodes is the half wave-length of the sound in air, $\frac{\lambda}{2}$. The half wave-length in the solid rod is its length L. These distances are traversed in the same time, and therefore

$$\frac{l}{L} = \frac{v}{V},$$

the ratio of the velocity of sound in the air and in the solid. Knowing the velocity of sound in air at the temperature of the experiment, the ratio gives the velocity in the solid, and this is connected with its coefficient of elasticity by the formula $V = \sqrt{\dfrac{e}{d}}$, d being density (117).

Kundt has by means of this apparatus confirmed the theory of von Helmholtz that:

1. The velocity of sound in a tube diminishes with the diameter, when the diameter is less than a quarter of the wave-length of the sound considered.

2. The diminution is greater for grave sounds than for acute ones.

Kundt has also verified the law that the velocity is independent of the pressure between 400 mm. and 1760 mm. of mercury.

He has also confirmed the law of the variation of velocity with temperature, viz., that velocity is proportional to $\sqrt{1 + at}$.

Similar apparatus serves to compare the velocity of sound in different gases.

157. Perturbations at the Extremities of Pipes (V., II, 132). — Experiments similar to those described confirm the indications of theory that the internodal distances are constant and equal to the half wave-length of the sound emitted, except at the open ends, and particularly the first segment next to the mouth-piece. Koenig found for the eighth sound of an open pipe, that is for the seventh overtone, the following distances in millimetres between antinodes, beginning at the mouth-piece:

173, 315, 320, 314, 316, 312, 309, 271.

The mean of the six middle ones is 314; the first is less

than this mean by 141, and the last by 43. The length of the pipe was 233 cms. and its breadth 12 cms.

The length of a pipe, open or closed, is less than the theoretical length of $\frac{\lambda}{2}$ or $\frac{\lambda}{4}$ for its fundamental tone. When the pipe gives a superior tone, the internodal distances are always $\frac{\lambda}{2}$ with the exception of the first and the last. These variations have been called perturbations at the extremities. The more important perturbation is the one at the mouth-piece.

At the free extremity of the open pipe, under the influence of the current of air traversing the pipe, the vibrating column is prolonged beyond the walls. The reflection from the external air is not then exactly at the plane through the extremity of the pipe, but a little further out, and the condensation at this plane of reflection is not rigorously zero. Moreover, there are probably multiple reflections at the ends of the pipe from the successive layers of air.

It is not, then, the length of the pipe itself, but this length augmented by a constant quantity l, which ought to be an exact multiple of the half wave-length for reënforcement by an open organ pipe. So also for a closed pipe the sounds energetically reënforced are those for which the length of the pipe, augmented by a constant quantity l', is an uneven multiple of a quarter of a wave-length. Wertheim found $l' = 0.746R$, where R is the radius of the pipe. Other investigators have found other values, and the whole difficulty remains to be resolved.

158. Beats due to Overtones (A. and B., 385). — Beats are produced not only between two notes nearly in unison, but between notes whose interval is *approximately* an octave, a major third, a fifth, and so on. These von Helmholtz attributes to the overtones associated with the fundamentals. Thus if two notes have vibration-frequencies n and $2n + 1$, then the first overtone of the lower will be due to $2n$ vibrations per second, and this will produce one beat per second with the higher note. So also if two notes are due to $2n + 1$ and $3n$ vibrations per second respectively, then the second overtone of the first will be due to $6n + 3$, and the first overtone of the second to $6n$ vibrations per second, giving three beats per second, though the interval is otherwise indistinguishable from a fifth. Combinations of such vibrations, obtained mechanically by Koenig, show periodic variations of amplitude corresponding with the beats.

Again, the interval between the fundamentals may be exact, but the overtones may be partial tones, and so not exact multiples of the fundamentals. Such is the case with tuning-forks, and beats are sometimes heard between their overtones of the same order.

159. The Quality of Sound (H., 106 (113); Z., 341; V., II, 292). — Two of the essential characteristics of musical sounds have already been considered, viz., pitch and loudness or intensity. But there is a third important difference between musical sounds, known as their *quality* or *timbre*. We easily recognize that one sound differs from another not only in being more acute or grave, louder or softer, but also in respect to the character of the sound itself. We have no difficulty in distinguishing the notes of the violin from those of the piano, even though they are of the same fundamental pitch and loudness. In the same way we learn to distinguish one voice from another in speech as well as in song, even when somewhat distorted in transmission by the telephone, or when reproduced by the phonograph. The same musical instrument may emit tones with marked differences depending upon the player; and even the untrained musical ear can readily distinguish between the character of the music produced by different instruments of the same class. The tones of a modern violin are far inferior to those emitted by an old Stradivarius, for example; and different players evoke different tones from a Stradivarius. All these differences, not assignable to pitch or loudness, are included under the term *quality*.

If we seek for the physical basis of the three characteristics of musical sounds, we know that pitch depends upon the wave-length, and loudness upon the amplitude of vibration; quality must therefore depend upon the only other physical difference between aërial sound-waves, viz., their vibrational form. By form is meant the law according to which the velocities or displacements of the air particles change from point to point along the path of the wave. This may be expressed either graphically or by means of a mathematical equation. Let the two upper curves in Fig. 96 represent two simple harmonic motions in the same medium with periods as two to three; the amplitudes are the same. Their resultant may be found by adding together corresponding ordinates with their proper sign. It is represented by the heavy line below. The complete vibrational form which is repeated over and over is not quite all shown in the figure. It recurs with three wave-periods of the first and two of the second component motions.

It will be remembered that according to Fourier's theorem, any periodic vibration admits of resolution into a fundamental or prime simple harmonic motion, and harmonics having frequencies represented by exact multiples of the prime. Any change in the vibrational form of such a wave, not affecting pitch or loudness, must then be due

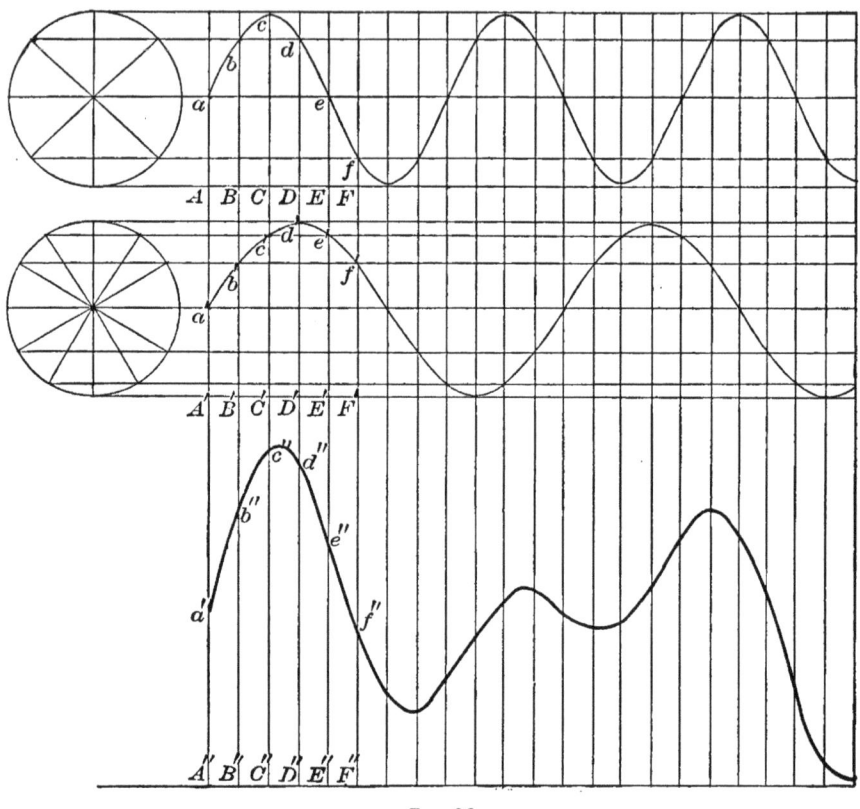

Fig. 96.

to some change in the components higher than the fundamental, or in the overtones.

If we consider partial tones arising from the several modes of vibration which a sonorous body may execute simultaneously, some of these may harmonize with the prime or be harmonic partials, while others are inharmonic. Those of a stretched string are in general harmonic, while those of bells, plates, and tuning-forks are inharmonic. But the motions of the air particles, conveying all these sounds to the ear, must be such as to represent all the component sounds, since at any point there can be at one instant only one definite density and a definite velocity in one direction. Each air particle has therefore impressed upon it the motions representing the several partial sounds, and its motion is the resultant of all of these. The ear possesses the marvellous property of analyzing this complex motion into its constituents and of thus selecting out the component tones which enter into the complex melange. Now the vibrational form of the complex resultant wave depends upon the presence of the overtones, which impress modifications upon the fundamental; and the musical quality of the tone is in the same way determined by the presence of the overtones associated with the fundamental. The conclusion of von Helmholtz, derived both from the analysis and the synthesis of musical sounds, is as follows : " The quality of the musical portion of a compound tone depends solely on the number and strength of the partial tones, not on their differences of phase."

The vibrational form of a sound-wave depends upon the number, the order, the relative intensity, and the relative phase of the overtones associated with the fundamental. But if the quality of a musical sound be defined by the number and the intensity of the simple sounds into which it is decomposable in accordance with Fourier's theorem, the phase does not appear to be able to intervene. Koenig has tried to show that phase difference among the overtones does produce a change in quality. His instruments for this purpose are called wave sirens. Physicists are divided in opinion respecting the validity of the conclusion which Koenig draws from his experiments. A change in phase certainly appears to produce a marked change in the quality of the resulting sound; but the objection is raised that Koenig's method of producing the component tones does not insure such simplicity of the sounds combined as one requires in order to admit the interpretation which he gives to these interesting phenomena.

160. Resultant Tones (V., II, 253 ; Z., 322 ; H., 229 (253) ; D., 438). — When two sufficiently intense sounds of frequencies n and n', differing less than an octave, are produced together, they give rise to a third sound, the vibration-rate of which is

$$N = n' - n.$$

These are called *resultant tones*. They were discovered by Sorge in 1740, and independently by Tartini in 1754; they are therefore often called Tartini's tones. Tartini gave to them the name of third tones.

If we sound together two forks c'' and g'', whose interval is a fifth and whose frequencies are 512 and 768, we obtain a distinct tone c' with a frequency of $768 - 512 = 256$. So also c'' and e'', with frequencies 512 and 640, give as a resultant tone $640 - 512 = 128$, or c.

These tones have been called *differential* tones by von Helmholtz, because their frequency is the difference of the frequencies of the two tones from which they arise. Other resultant tones, called *summational tones*, were discovered by von Helmholtz. Their frequency is the sum of the frequencies of the tones producing them, or

$$N = n' + n.$$

Thus c' and g', frequencies 256 and 384, will produce together e'' of 640 vibrations per second, and

$$256 + 384 = 640.$$

Summational tones are much feebler than differential tones — a fortunate circumstance, since they are mostly inharmonic.

Resultant or combinational tones may be attributed to two causes. The first one is in the external air, and the other has its origin in the drum of the ear.

When two disturbances from separate sources are superposed, we have assumed that the amplitude of the resultant disturbance at any point is the sum of the two separate disturbances. This is true only so long as the displacements are very small. The amplitude of the compound oscillation for greater disturbances falls short of the sum of the amplitudes of the components. This is equivalent to the introduction of another vibration whose frequency is that of the differential tone. Particular instruments give powerful combinational tones when the same mass of air is violently agitated by two simple tones simultaneously. This is the case with a siren in which two or more rows of holes are blown from the same wind-chest; and with the harmonium, in which openings, closed and opened rhythmically by the tongues or reeds, communicate with a common wind-chest. This objective portion of combinational tones can be reënforced by resonators.

But the vibrations corresponding to combinational tones may exist in the parts of the ear without any objective existence in the external air. The construction of the external ear is peculiarly favorable to the generation of combinational tones. Its external convex radial fibres undergo a greater change of tension when they oscillate with a moderate amplitude inwards than when the oscillation is outwards. "Under these circumstances deviations from the simple superposition of vibrations arise for very much smaller amplitudes than is the case where the vibrating

body is symmetrically constructed on both sides." Sounds
so produced cannot be reënforced by resonators.

**161. Dissonance due to Beats (V., II, 306 ; Z., 422 ;
H., 251 (277)).** — The theory of von Helmholtz to explain
dissonance proceeds from the fact that beats produce in the
nerves of audition an intermittent excitation, and this is
disagreeable as are all excitations of this kind, like that
produced by a fluttering light. According to von Helmholtz beats become most disagreeable at 33 per second. At
132 per second they cease to be perceptible. Further, the
distinctness of the beats and the harshness of the interval
do not depend solely on the absolute difference in the
number of vibrations ; they are dependent also upon the
ratio of these numbers, or, as we commonly say, the magnitude of the interval. If we could disregard their magnitude,
all the following intervals, which give 33 beats each, would
be equally rough :

The semitone . . . . . b', 495 — c'', 528.
The whole tone. . . . . c', 264 — d', 297.
The minor third . . . . e, 165 — g, 198.
The major third . . . . c, 132 — e, 165.
The fourth . . . . . . G, 99 — c, 132.
The fifth . . . . C, 66 — G, 99.

But we find the deeper intervals more and more free
from roughness. If the two sounds are sufficiently separated in the musical scale, the audition fibres simultaneously
affected by the two sounds vibrate too feebly for the beats
to be appreciable. For a given frequency of beats, therefore, the harshness of the effect increases with the nearness
of the notes on the musical scale.

Thus far we have supposed the sounds producing dissonance are two simple tones. But musical sounds are always accompanied by harmonic partials, the importance of which is apparent upon inspection of the several intervals with their partials below:

Unison 1 2 3 4 5 6 7 8 9 10
　　　　　　　　　　 | | | | | | | | | |
　　　　　　　　　　 1 2 3 4 5 6 7 8 9 10

Octave 1 2 3 4 5 6 7 8 9 10
　　　　　　　　　 | 　 | 　 | 　 | 　 |
　　　　　　　　　 2 　 4 　 6 　 8 　 10

Twelfth 1 2 3 4 5 6 7 8 9 10
　　　　　　　　　 　 | 　 　 | 　 　 |
　　　　　　　　　 　 3 　 　 6 　 　 9

Fifth 2 4 6 8 10 12 14 16 18 20
　　　　　　　 　 　 | 　 　 | 　 　 |
　　　　　　　 　 3 6 9 12 15 18

Fourth . . . 3 6 9 12 15 18 21 24 27 30
　　　　　　　 　 　 | 　 　 　 |
　　　　　　　 4 8 12 16 20 24 28

Major sixth . . 3 6 9 12 15 18 21 24 27 30
　　　　　　　　 　 | 　 　 　 |
　　　　　　　　 5 10 15 20 25 30

Major third . . 4 8 12 16 20 24 28 32 36 40
　　　　　　　　 　 | 　 　 　 |
　　　　　　　　 5 10 15 20 25 30 35 40

The larger the number of pairs of upper partials in any interval near enough to each other on the scale to produce distinct beats, the greater will be the dissonance of the interval. If such beating pairs are altogether wanting, or have so little intensity that their effect is negligible, the interval is a consonant one.

Four distinct groups can be readily distinguished.

1. *Unison*, the *octave*, and the *twelfth*, to which also may be added the *double octave*. The consonance here is

perfect. The prime and partials of the upper tone all coincide with partials of the lower.

2. The *fifth* and the *fourth*. These may also be regarded as perfectly consonant, because they may be used in all parts of the scale without any disturbance of the harmony. The fourth is the less perfect consonance. It owes its superiority in musical practice simply to its being the defect of a fifth from an octave.

3. The *major sixth* and *major third*. These are called *medial consonances*. They were considered imperfect consonances by old writers on harmony. The coincidences among the upper partials are few in number, and the beating pairs are more numerous than in the preceding intervals.

4. The *minor third* and the *minor sixth*. These are *imperfect consonances*. Coincident pairs of partials are almost or quite wanting, and the pairs near enough to beat are either the semitone 16 : 15 or the semitone 25 : 24. These partials, on which the definition of the interval depends, are often absent, or the imperfection in the interval produced by them is slight. (Not shown on preceding page.)

162. The Musical Importance of Resultant Tones (V., II, 256). — Resultant tones play an important rôle in music by reason of the influence which they exert on consonance.

Consider, for example, a perfect major chord in which the vibration-frequencies have the relation

$$4, \quad 5, \quad 6, \quad 8,$$

or

$$1, \quad \frac{5}{4}, \quad \frac{3}{2}, \quad 2.$$

The resultant difference tones are

$$\frac{1}{4}, \quad \frac{1}{2}, \quad \frac{3}{4}, \quad 1,$$

or the double lower octave of the fundamental, the lower octave of the same, the octave below the fifth, and the fundamental sound; all of these reënforce the primary sounds and particularly the fundamental. The differential tones all strengthen the consonance.

To illustrate numerically, let the major chord consist of c', e', g', c''. These ˙with their differential tones may then be written as follows:

C	c	g	c'	e'	g'	c''
64	128	192	256	320	384	512
1×64	2×64	3×64	4×64	5×64	6×64	8×64

The entire series consists of C with its harmonic partials up to the eighth in order, with the exception of 7×64, which would mar the consonance. The differential tones contain no discordant elements.

In the perfect minor chord the frequencies are

$$10, \ 12, \ 15, \ 20,$$

or

$$1, \ \frac{6}{5}, \ \frac{3}{2}, \ 2.$$

The differences are

$$\frac{1}{5}, \ \frac{3}{10}, \ \frac{1}{2}, \ \frac{4}{5}, \ 1.$$

These are the double octave below the grave · major third; the double lower octave of the minor third; the octave below the fundamental; the grave major third; and the fundamental. A new sound appears here, the grave major third, which is dissonant with the fifth. This explains the indecisive, pathetic character of the minor chord.

163. Musical Pitch (D., 385; Z., 75). — A progressive change in the absolute pitch of the notes of the arbitrary scale employed by musicians has been going on for many years. Attempts have been made in recent times to arrest this upward tendency, and with some success.

In the time of Handel the middle A (a') of the piano, or the second string of the violin, was made by 424 vibrations a second; while the organ pitch in England in the middle of the eighteenth century was as low as $a' = 388$ vibrations. At Paris in 1700 middle A was 405; later on, 425; in 1855, 440; and in 1857, 448 vibrations per second.[1]

Quite recently the pitch of the Théâtre de la Scala in Milan was 451.5, and that of the Covent Garden Theatre, London, 455. Modern concert pitch has risen as high as 460 for a'. The sharping of the notes gives a certain brilliancy to the music of orchestral instruments, but makes a great demand upon the powers of concert singers, particularly as much modern music is written upon high notes.

"It was necessary to find a remedy for so grave an inconvenience, and therefore an international commission fixed as the *normal pitch* a tuning-fork giving 435 vibrations per second." This corresponds with the recommendation of the Paris Academy of Sciences.

The German Society at Stuttgart in 1834 recommended $a' = 440$ as the standard pitch. This makes $c'' = 528$ vibrations per second. Since 528 is 22 times 24, by multiplying the series representing the vibration-rates of the perfect diatonic scale, 24, 27, 30, 32, 36, 40, 45, 48, by 11, we obtain the vibration-frequencies of the notes of the natural gamut of C corresponding to this standard of pitch.

For scientific purposes c' is taken as 256 vibrations per second. Since this is the eighth power of 2, any power of 2 will express the vibration-frequency of C in some octave, according to this standard. On the same standard a' is $426\frac{2}{3}$.

---

[1] Blaserna's *Theory of Sound*, p. 70.

164. Limits of Pitch employed in Music (Bl., 67; B., 243). — In the modern pianoforte of seven octaves the bass A corresponds to about 27.5 vibrations per second, the highest A to 3480. Taking into account the slight variations from standards in tuning, the notes employed on the piano of seven octaves range between 27 and 3500 vibrations per second. Some pianos go as high as the seventh C with 4200 vibrations per second, but such high notes are shrill and lacking in sweetness. The practical limits are about 27 and 4000.

On the organ the deepest note is the C of 16 vibrations per second given by the 33-foot open pipe. Its wave-length in air at normal temperatures is about $\frac{344}{16} = 21.5$ metres or 70.5 feet. The highest note is the same as the highest A of the piano, which is the third octave above $a' = 435$. It is made therefore by $435 \times 2^3 = 3480$ vibrations per second.

The well-developed voice of a singer embraces about two octaves. The voice of woman is represented by about twice as many vibrations per second as that of man. The lowest note of the voice, not including certain exceptional cases, is C of 65 vibrations. The entire range is included between this note and c''' of 1044 vibrations per second, the two higher octaves belonging to woman. The highest exceptional voice appears to be that of Bastardella, whom Mozart heard at Parma in 1770. Bastardella's voice had a range of three and a half octaves and went up nearly to 2000 vibrations.

The limits of hearing far exceed those of the voice or of music. They are about 16 for the lowest note, and from 38,000 to 40,000 for the highest. The limits are somewhat different for different persons; but we may say that sonorous vibrations lie between 16 and 40,000 per second.

LIGHT.

CHAPTER VIII.

NATURE AND PROPAGATION OF LIGHT.

165. Nature of Light. — The sensation of light is due to a mechanical action upon the extension of the optic nerve, forming the sensitive surface of the retina. The undulatory theory of light, now universally accepted, assigns this action to a disturbance propagated from its source by a wave-motion in the universal medium called the *ether.* This process of *radiation*, as it is called, is clearly a process of transference of energy through the ether; and this transfer is accomplished by periodic disturbances in the medium which follow the laws of wave-motion. These disturbances, according to the theory of Maxwell, which has been confirmed by the remarkable experiments of Hertz, are electro-magnetic phenomena. Light objectively " now takes its place alongside of electric phenomena, as but one of the forms of energy associated with that wonderful kind of matter provisionally called the ether " (Tait).

The existence of the ether is a necessary consequence of Roemer's discovery in 1676 of the finite speed of light. For the transmission of light is the transmission of energy; and a medium of transmission is a necessary postulate as the repository of this energy during the time of transmission.

The earth receives energy from the sun; and as something over eight minutes are consumed in its transit across the intervening space, we are forced to seek for the vehicle by which it is conveyed. Only two methods of the propagation of energy are known to us, and no other method seems now possible or conceivable. Energy, according to our present knowledge, is always associated with matter, so that matter has been defined as the vehicle of energy.

One of the methods by which matter conveys energy to a body at a distance is that of a projectile. The bullet carries its energy of motion from the gun to the mark across the intervening space. This is Newton's corpuscular theory of light. He imagined the light-giving body projecting minute particles, or corpuscles, through space; and that these, entering the eye, excited vision by impact upon the retina.

The other method of propagation requires a continuous medium, and the energy is handed along from particle to particle as an undulation. In this way energy is conveyed in sound and by water-waves across the surface of the sea. In the first method the energy remains with the matter transmitting it from start to finish; in the second it is passed along from particle to particle by the series of operations which transmit the wave-motion. According to this latter theory a luminous body is the centre or source of a disturbance in the ether, which propagates it in waves through space. They travel with the velocity of light, and entering the eye excite the sense of vision.

Now both of these methods involve the element of time, and conversely the existence of the time element appears to limit the transmission to these two methods. But Newton's corpuscular theory failed because of its complexity and the crucial test applied to it by Foucault (Art. 172);

there remains then only the undulatory theory and the ethereal medium.

Physical optics consists in a study of wave-motion propagated through the ether in accordance with the principle of Huyghens. It is an inquiry into the physical processes concerned in the transmission of light and in the phenomena of reflection, refraction, interference, and dispersion. It seeks to account for them entirely by the application of dynamical principles. *Geometrical optics*, on the other hand, is confined to the exposition of those facts which can be studied by the aid of the simple geometrical considerations involved in the laws of reflection, refraction, and the transmission through isotropic media in right lines or *rays*, without any inquiry into the nature of light itself.

In this elementary treatment of the subject it will best suit our purpose to adhere exclusively neither to the one method nor to the other, but to make use of either the physical or the geometrical method according as it may best serve for the simple exposition of fundamental principles.

166. The Rectilinear Propagation of Light (V., II, 311 ; L., 14 ; T., 24). — It is a fact of common experience that in a uniform medium, possessing the same properties in all directions, light travels in straight lines or rays. If a source of light, as a candle, is concealed from the eye by an opaque screen, it suffices to move the screen so as to uncover the right line connecting the candle and the eye in order that the light may be seen. So no objects can be seen through a straight tube except those situated in the direction of the axis produced.

But rectilinear propagation is confined to media technically described as both homogeneous and isotropic. In other words the wave-surfaces must be truly spherical about the source as a centre. In such a medium the disturbance at any point is due almost entirely to the motion transmitted along a perpendicular dropped upon the wave-surface in some prior position. The secondary waves transmitted from all other points of the surface, as new centres of disturbance, neutralize one another along this perpendicular on account of the excessive minuteness of the wave-lengths of all radiations exciting vision.[1]

But in media of unequal speed of transmission in different directions the wave-surfaces are not spherical, a ray of light is no longer necessarily at right angles to the wave-surface, and light may be transmitted in curved lines if the properties of the media vary from point to point.

If we look along a hot bar or over a heated surface objects beyond appear distorted. So when the sun shines on a hot stove, the rising currents of heated air produce the appearance of images or shadows on a white wall beyond. The air is irregularly expanded, the medium is not homogeneous, and the light no longer moves in straight lines. To this lack of homogeneity are due the phenomena of mirage, the duplication of images of a distant object seen through an atmosphere unequally heated above and below, and the twinkling of the stars. So the rectilinear propagation of light is limited to media having uniform optical properties in all directions and without variations from point to point.

167. Images produced by Small Apertures (T., 30; V., II, 315). — If light from a luminous surface pass through a small opening of any shape in an opaque screen and fall upon a white surface at some distance, it will form

---

[1] Preston's *Theory of Light*, p. 54.

an inverted image of the source (Fig. 97). Not only is this image *EF* inverted, but it is *perverted*. By making a

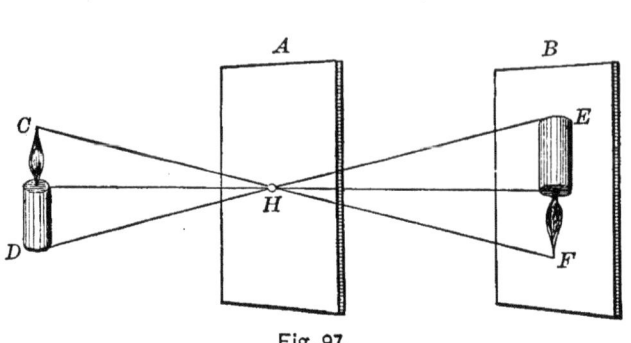

Fig. 97.

small opening in the shutter of a darkened room an image of outside objects, brightly illuminated by the sun, may be obtained upon a white screen held at a suitable distance. Not only is there inversion, shown by objects being depicted lower the higher they are, but if the image, viewed from the side of the screen toward the opening, be imagined turned round in its own plane so as to make it erect, it will be found that the right side of the object, as we look toward it, corresponds with the left of the image. If the screen is translucent and the image is viewed from behind, it will then be inverted, but not perverted. The right hand may be considered the perverted image of the left. An image in a plane mirror is perverted, but not inverted.

Each point of the object (Fig. 97) is the vertex of a cone of rays passing through the aperture and forming an image of it on the screen. These images will be symmetrically placed with reference to the points emitting the light, and consequently by their superposition will form a figure of the same outlines as the luminous object. Since a smaller number of these images of the opening overlap near the edges of the image on the screen, the edges of the picture will be less bright than the other portions. Moreover, the sharpness of the image diminishes as the opening is made larger; all the images of the opening become larger, and with a large opening their superposition no longer produces a picture resembling the object but rather the opening. The insertion of a converging lens in the opening not only permits of a larger aperture, thus increasing the illumination, but improves the definition of the image by causing the light emanating from any point of the object to come to a focus at the corresponding point of the image with minimum overlapping of neighboring images.

A particular case of the preceding phenomenon is furnished by the light of the sun, sifting through the foliage of trees and tracing circular or elliptical images on the ground. During solar eclipses these spots change into crescents, which are the more pronounced as the eclipse is more complete.

168. Theory of Shadows (T., 26; L., 15; V., II, 311). — An optical shadow is the region from which light is wholly or in part cut off by an opaque body. If the body is in the vicinity of a luminous point its illuminated area is limited by the curve of contact of a cone starting from the point and circumscribed about the opaque body *AB* (Fig. 98). The portion of the space within this cone of shadow and beyond the curve of contact is entirely in obscurity.

But when the luminous source has sensible dimensions, then outside of the total shadow or *umbra*, and surrounding it, is a region called the partial shadow or

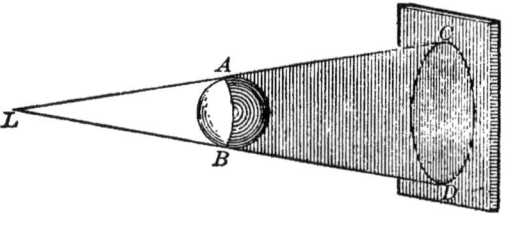

Fig. 98.

penumbra. This is limited by the double cone with its apex between the luminous and the opaque bodies (Fig. 99).

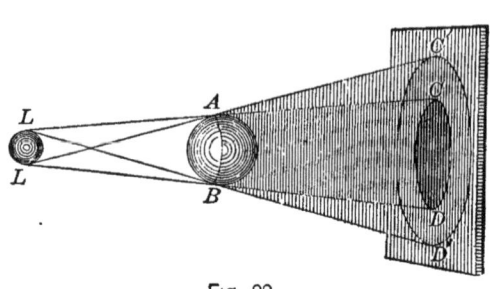

Fig. 99.

This region of partial shadow forms the transition from complete obscurity to the full light. The smaller the angular dimensions of the source and the nearer the screen CD to the opaque body, the narrower will be the penumbra. Only a portion of the luminous body is visible to an eye situated within the penumbra.

Solar eclipses are produced by some portion of the earth's surface passing through the shadow of the moon. Since the moon is smaller than the sun its shadow is a limited cone; and the apex of this shadow cone sometimes reaches the earth, when new moon occurs near one of the lunar nodes; and sometimes it falls short of it, the mean length of the lunar shadow being less than the mean distance of the moon from the earth. If the shadow cone reaches the earth the eclipse is total for all points within the umbra; within the penumbra the eclipse is partial; if only the prolongation of the shadow cone encounters the earth the eclipse is annular for all points touched successively by the axis of the shadow.

169. Speed of Light from the Eclipse of Jupiter's Satellites (T., 43; B., 397). — The eclipses of the inner satellite of Jupiter occur at average intervals of 42h. 28m. 36s. It moves much faster than our moon, so that the eclipses appear from the earth to take place quite suddenly, though, since it is really a gradual phenomenon, any one observation is doubtful to half a minute.

In 1676 Roemer, a Danish astronomer, an observer at the time in the Paris Observatory, discovered that the observed eclipses differ systematically from the computed times. When the earth is receding from Jupiter the interval between two successive eclipses is longer than the mean, and the more rapid the recession the greater the excess. The reverse is true as the earth approaches Jupiter. Let *EE'* (Fig. 100) represent the earth's orbit and the large circle *JJ'* the orbit of Jupiter. Then while the earth moves from *E* through *E'* to *E''*,

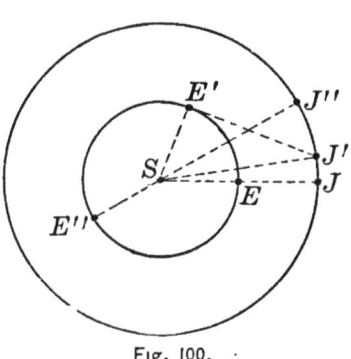

Fig. 100.

or from opposition to conjunction, the eclipse interval is longer than the mean; from *E''* around to opposition again it is shorter than the mean. The sum of all these excesses is 16m. 38s., or 998 seconds.

Roemer inferred that the speed of light is finite, so that the longer interval between two eclipses when the earth is receding from Jupiter is due to the added distance which light must travel to reach the earth. This interval will be the greatest at *E'* where the earth is receding directly from the planet. The sum of the excesses is the time required by light to travel across the earth's orbit. If the diameter be 299,000,000 kilometres, the speed of light will be $\dfrac{299,000,000}{998} = 299,600$ kilometres per second.

Roemer's original suggestion was rejected by most astronomers for more than fifty years, and was not accepted till long after his death, when Bradley's discovery of the aberration of light confirmed the correctness of Roemer's views (Young's Astronomy).

170. Bradley's Method from the Aberration of Light (T., 44; P., 10). — Aberration is the displacement in the apparent position of a star resulting from the composition of the motion of light with the motion of the earth. It was discovered by Bradley, afterwards Astronomer Royal of England, in 1726. If an observer carrying an umbrella walk rapidly, while the rain falls vertically, he must tilt his umbrella forward if he would protect himself; for to him the rain does not appear to come in the same direction as to one standing still. He must incline his umbrella forward at an angle a with the vertical so that $\tan a = \dfrac{u}{V}$, where V is the velocity of the falling drops and u the speed with which he is walking.

Suppose the wind blowing directly against the side of a moving vessel, with a velocity V, and that the vessel move with a velocity u. The motion of the steamer produces an apparent wind blowing toward the stern; and this combined with the real wind causes, to an observer on the vessel, an apparent shifting forward of the point from which the wind comes by an angle a of which the tangent is $\dfrac{u}{V}$ as before. It was an observation of this kind that gave Bradley the clue to the explanation of the apparent displacement which he had observed in the position of stars when viewed from opposite sides of the earth's orbit. The real speed of light must be combined with one equal and opposite to that of the earth in its orbit in order to give the apparent direction from which the light comes, or the apparent position of a star.

Let CA (Fig. 101) represent the velocity of light V, and let AB be the relative magnitude and direction of the orbital velocity u of the earth. Then will CAD be the angle of aberration. This angle will be greatest when the motion of the earth is at right angles to the direction of the star. Then $\tan a = \dfrac{u}{V}$. The angle of aberration is known to be almost exactly $20''.5$. Since the tangent of this angle is about $\frac{1}{10000}$, it follows that the speed of light is about 10,000 times the earth's orbital velocity; and as the latter is about 30 kilometres a second, the speed of light is about 300,000 kilometres a second.

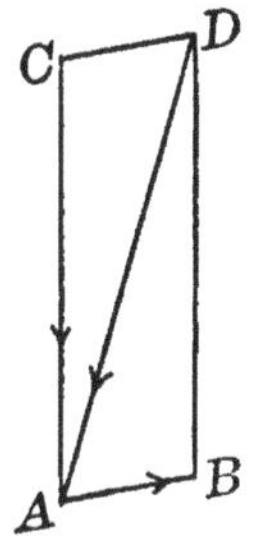

Fig. 101.

This explanation of aberration is really founded on the corpuscular or projectile theory of light. In the wave-theory it presents difficulties that have not yet been surmounted. It is nevertheless true that the ratio between the speed of light and the constant of aberration is about $\frac{1}{10000}$.

Both of these astronomical methods of measuring the speed of light depend for their final result upon an exact knowledge of the earth's distance from the sun. But it is probable that the most accurate determinations of this distance are to be made by reversing the order of computations. Given the speed of light, independently determined by physical processes, the constant of aberration, or the retardation of the eclipses of Jupiter's satellites, furnishes a measure of the sun's distance.

171. Fizeau's Direct Method (P., 401; B., 399). — The first direct measurement of the speed of light over limited terrestrial distances was made by Fizeau in 1849.

His method depends upon the eclipse of a source of light by means of a rapidly rotating toothed wheel. The principle is therefore analogous to Roemer's observations on the satellites of Jupiter.

A beam of light from a source *S* (Fig. 102) passes through a collimator in the side of a telescope; and, after reflection from a parallel plate of glass set at an angle of 45°, it comes to a focus at *F*, which is the principal

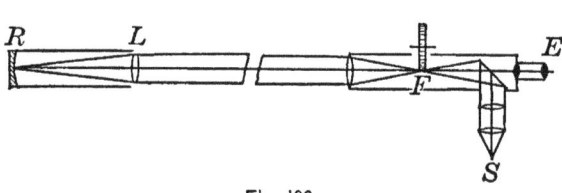

Fig 102

focus of the object glass of the telescope. The light diverging from *F* emerges from the object glass in a parallel beam; and, after traversing a distance of three or four miles, it falls upon a lens *L*, which brings it to a focus on the surface of the concave mirror *R*, having its centre of curvature at the lens. The beam of light therefore returns along its former path, enters the telescope and falls upon the inclined plate of glass; part of it is reflected and a part transmitted; and, after traversing the eyepiece *E*, it enters the eye of the observer, producing the appearance of a bright star at *F*.

A toothed wheel is so placed that as it rotates the beam of light is alternately intercepted and allowed to pass between the teeth. Suppose the angular breadth of the teeth is equal to the width of the spaces between them.

If now the speed of light were infinite, the illumination of the star would not be affected by the rotation of the wheel; but if it is finite, a rate of rotation may be found, such that the light going out through a space will be intercepted on its return from the distant station by a tooth, and there will be complete extinction.

What occurs in the experiment is at first the appearance of a bright star, which gradually diminishes in brightness as the speed of rotation increases, till at a fixed speed it is entirely extinguished; if the speed of rotation continues to increase the star reappears, reaches its former maximum brightness, again fades out, and is eclipsed when the speed of rotation is three times that required for the first eclipse.

It was found very difficult to decide upon the exact speed required to produce a complete eclipse. The light was much weakened by successive reflections, so that the star was faint even at its greatest brightness. It was rendered less distinct by the diffused light in the interior of the telescope, caused by reflection from the teeth of the wheel. Fizeau found the first eclipse at a speed of 12.6 revolutions a second. There were 720 teeth, or 1440 divisions of the wheel. Hence the time required for a tooth to take the place of a space was $\frac{1}{12.6} \times \frac{1}{1440} = \frac{1}{18144}$ of a second. The double distance between the telescope and the reflector was 17.326 kilometres. Hence the speed of light as deduced from this experiment is 17.326 × 18144 = 314,363 kilometres.

In 1874 Cornu repeated Fizeau's experiment with greatly improved apparatus, the improvements consisting chiefly in better methods of recording the exact speed of rotation of the wheel at any observed phase of the eclipse. No difference in result was obtained with different sources of light. Cornu's final result for the speed of light was 300,330 kilometres in air. To obtain the speed in outer space, free from ponderable matter, this result must be multiplied by the index of refraction of air (187) which gives 300,400 kilometres per second in a vacuum.

**172. Michelson's Modification of Foucault's Method
(A. and B., 434; P., 409; T., 49).** — In 1850 Foucault
described a method of measuring the speed of light, founded
upon the measurement of time by means of the rotating
mirror employed by Wheatstone in 1834 to measure the
speed of electricity. Foucault's method was intended
primarily to compare the speed of light through air and
water as a crucial test between the emission and the undulatory theory of transmission (188). A focusing lens
was placed between the source and a plane mirror which
reflected the beam to a concave mirror. The centre of the
concave mirror was at the plane mirror, which was mounted
to rotate around an axis in its own plane. The distance
between the two mirrors was originally only four metres.
If the plane mirror were standing still, then in a certain
position the light reflected to the concave mirror would retrace its path, and could be observed by means of an eyepiece at the source after reflection from an obliquely set
plate of plane glass. But if the plane mirror were rotated,
then the beam of light, after travelling to the concave
mirror and back, would find the plane mirror in a slightly
different angular position, and during the remainder of its
return passage to the eyepiece it would pass by a slightly
different path. The divergence of the two positions was
measured by means of a micrometer in the eyepiece. This
divergence, together with other constants of the apparatus
and the speed of rotation of the mirror, gave a measure of
the speed of light.

But in Foucault's arrangement the deflection of the
beam was too small to be measured with the requisite
accuracy, being but a fraction of a millimetre. Michelson's
modification was designed to increase this displacement. Its
most important feature was the placing of the lens between

the two mirrors, m and m'. Moreover, the fixed mirror, which was plane, was placed at a distance of 605 metres from the rotating mirror. The displacement of the beam was increased to 133 mm., or about 200 times that obtained by Foucault.

An outline of Michelson's arrangement is shown in Fig.

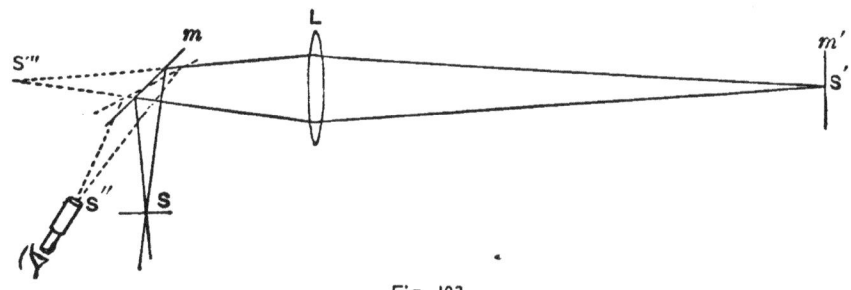

Fig. 103.

103. At S is a narrow slit, m is the revolving mirror, L the lens, and m' the fixed mirror. Light from a source behind S passes through a slit, falls on m, is reflected to m' whenever m is in a suitable position, and forms an image at S'. Light reflected from m' through the lens L comes to a focus at S. The image of S in the first mirror and S' are conjugate foci of the lens (196).

When m rotates rapidly its position will change while the light travels from m to m' and back, and the reflected beam will accordingly be displaced to some position S'' in the direction of the rotation. As finally arranged the revolving mirror was 8.58 metres from the slit, and the distance between the two mirrors was 605 metres. With a speed of 257 revolutions a second, the observed deviation was 113 millimetres.

Michelson's final result for the speed of light in a vacuum was 299,853 $\pm$ 50 kilometres a second. By a similar method, and a distance between the mirrors of 3720 metres, Newcomb in 1882 obtained 299,860 ± 30 kilometres.

When a tube with glass ends and filled with water is interposed between L and m' the displacement of the image is increased, demonstrating that light travels slower through water than through air, as the undulatory theory requires.

PROBLEMS.

1. A candle and its image made by a small opening (167) are at distances 50 and 75 cms. respectively from the opening. Find their relative size.

2. If the intensity of light varies inversely as the square of the distance from the source, find the relative quantities of light emitted by a gas jet and a candle when they are 5 and 1.2 metres distant respectively from the photometer disc which they illuminate equally.

3. Assuming the velocity of light to be 300,000 kilometres a second, and the wave-length of sodium light 5890×10^{-10} metre, what is the frequency of vibration of the light of burning sodium?

4. If a photograph be taken in one ten-thousandth of a second by light whose wave-length is 4000×10^{-10} metre, what is the length of the beam falling on the plate, and how many waves are impressed on it?

CHAPTER IX.

REFLECTION AND REFRACTION.

173. Law of Reflection (T., 53; L., 26). — When a beam of light, propagated in one medium, encounters a second, a division of the light generally takes place between the two media. One portion enters the second medium, and follows one or two new paths; the remainder travels backward in the first medium, and is in general further separated into two portions, the relative intensities of which depend upon the polish of the surface of separation between the two media. If this surface were perfectly polished, all the reflected light would be confined to a single direction, and the reflecting surface itself would be invisible. On the contrary, an entirely rough, unpolished surface reflects the light irregularly in all directions. This is called diffused light. In general both processes go on together. Non-luminous objects become visible by the diffusely reflected light. In one important class of cases the reflection is *total*.

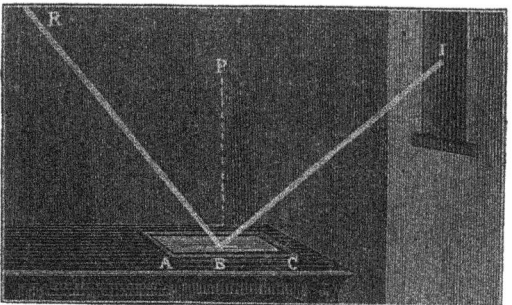

Fig 104.

When a beam of light falls on a polished surface *A C* (Fig. 104), or on the surface of a liquid, a large part of it is reflected in a

definite direction, *BR*. The line *PB* - is normal to the
reflecting surface at the point of incidence, the angle *IBP*
is the *angle of incidence*, and *PBR* is the *angle of reflection*.

The law of reflection is as follows :

*The incident and reflected rays are in the same plane
normal to the reflecting surface, and make equal angles with
the normal at the point of incidence.*

The two media considered above are not necessarily
different in their chemical composition, but of different
optical density. In special cases layers of water or of air
of different temperatures give surfaces of separation at
which reflection and refraction take place, as in the case of
aërial echoes in sound. When the transition in the density
of a substance in the same molecular state is gradual, the
reflection is slight, and the path of the refracted ray
becomes curved in the non-homogeneous medium.

It has been known from ancient times, and can be demonstrated mathematically (V., II, 319), that if a ray pass
from one point to another, after reflection from a fixed
surface, its whole path, touching the reflecting surface,
is the shortest that can be traced from the one point to
the other when the angles of incidence and reflection are
equal to each other.

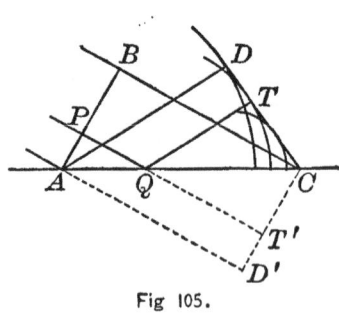

Fig 105.

**174. Law of Reflection de-
duced from the Undulatory Theory (T., 174; P., 66).** — Let *AB*
(Fig. 105) be a plane wave-front,
and *AC* the reflecting surface.
Each portion of this surface, as the
wave reaches it, becomes a new
centre for a diverging wave in the
first medium. Then during the time
required for the disturbance at *B* to reach *C*, the disturb-

ance at A has spread back into the medium as a spherical wave, whose radius is AD' equal to BC. Its section is a circle drawn from A as a centre. During the same time the light from P, which would have proceeded to T' if there had been no obstruction, has reached Q, and has developed into a spherical wave of radius QT equal to QT', whose intersection with the plane of the paper is the circle through T. All the circles which can be drawn in this manner intersect in the straight line CD. This is, therefore, a section of the reflected wave-front.

The triangle ABC equals $AD'C$ equals ADC. Therefore the angle BAC, which is the angle of incidence (135), is equal to ACD, the angle of reflection. Moreover, the incident and refracted rays lie in the plane containing the normal line. The undulatory theory, therefore, furnishes an adequate explanation of the law of reflection.

175. Images in a Plane Mirror (T., 59 ; V., II, 323). — An object is rendered visible by the rays diverging from it and entering the eye. Hence a pencil of diverging rays, coming from a point or diverging as if they came from a point, will convey to the eye the impression of a luminous source at that point. Any perception conveyed through the eye is referred directly outward in the direction in which the light enters the eye. The eye can give us information only about the stimulus which reaches it ; it furnishes no direct evidence of the source from which the stimulus comes, nor of the manner in which it manages to reach it. An *image* is therefore a point or a series of points from which a diverging pencil of rays comes or appears to come.

From the general law of reflection it results that all rays emanating from the point A (Fig. 106), and falling upon the plane mirror *MN*, have after reflection the same direction as if they came from the point *A′*, symmetrically situated with respect to the mirror. In couse-

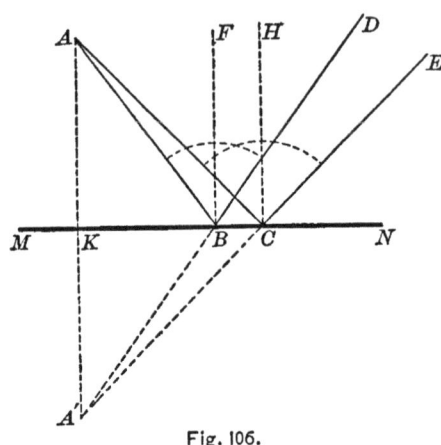

Fig. 106.

quence an eye placed at *DE* will be affected by these rays as if they came directly from *A′*. The point *A′* is accordingly called the image of *A* in the mirror *MN*. It is, moreover, called a *virtual* image to indicate that it is formed by the concourse of the prolongation of the rays, and not by the rays themselves. An image does not really exist at *A′*.

(*a*) The image of a point in a plane mirror lies on a perpendicular drawn from the point to the mirror, and as far behind the mirror as the object is in front. This proposition is readily demonstrated as follows:

Since the plane of the incident and reflected rays contains the normal, the ray *BD* projected backward must intersect the normal *AK* in some point *A′*. Then the angle $KAB = ABF = FBD = KA'B$. Hence the two right triangles, *AKB* and *A′KB*, are equiangular; and since the side *KB* is common to the two, they are equal to each other. *A′K* is therefore equal to *AK*. But since *ABBD* is any reflected ray from *A*, all the rays after reflection will diverge from *A′*, which is as far behind the reflecting surface as *A* is in front of it.

(*b*) The divergence of the rays after reflection is the same as before reflection. Hence the image is not distorted. Let AB and AC be any two rays from A. Then the two triangles ABC and $A'BC$ are equal to each other; for from the last pair of equal triangles AB is equal to $A'B$; in the same way AC is equal to $A'C$; and since the two triangles have their third side BC in common, they are equal to each other, and the angle BAC equals the angle $BA'C$. But the former is the angle of divergence before reflection and the latter the divergence after reflection.

176. Path of the Rays to the Eye. — The image of an object is the assemblage of the images of its points. The image may therefore be found by dropping perpendiculars from its several points upon the mirror and producing them till their length is doubled. Thus $A'B'$ is the image of AB in the mirror MN (Fig. 107).

Let E and E' represent two different observers. To find the path of the rays entering the eye at E, connect A' and B' by straight lines to E. From their intersections with the mirror draw lines to the object, A and B. Then the

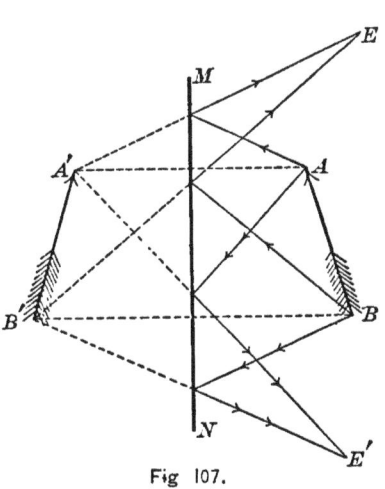

Fig 107.

full lines in front of the mirror represent the path of the rays from A and B, which give the image at A' and B'. The rays for the eye at E' are found in the same manner.

177. The Deviation by Successive Reflection from two Mirrors. — The deviation of a ray of light produced by two reflections from a pair of plane mirrors is twice the angle between the mirrors.

Let the ray be successively reflected from the two mirrors at E and F (Fig. 108). Then the deviation is the angle ϕ. We have

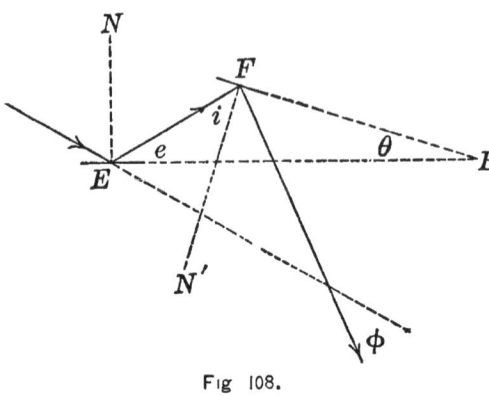

Fig 108.

$$\phi = 180° - 2\,(e + i).$$
$$\theta = 90° - (e + i).$$

Doubling the second equation and subtracting from the first,

$$\phi - 2\theta = 0.$$

But θ is the angle between the two mirrors.

178. Images of Images (T., 63; A. and B., 410; V., II, 353). — When light is reflected successively from two plane mirrors, the image in the first becomes the object for the second mirror, and the second image is found in precisely the same manner as the first one. So the second image may serve as the object for a third image, and so on, since in each case the light approaches either mirror as if it came from the next preceding virtual image in the other one.

Let O be a luminous point between the two mirrors AB and AC (Fig. 109). The first image in AB is found by drawing the perpendicular Ob and making the distances of b and O from the mirror equal. Then b is in front of the mirror AC, and its image c' is determined by the perpendicular bc'; c' is in front of the mirror AB and has its image at b''. But b'' is behind the plane of both mirrors and there is therefore no image of it.

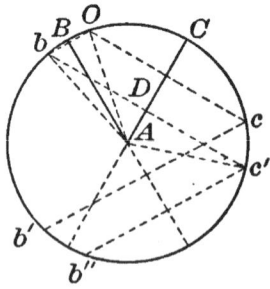

Fig. 109.

In the same way may be found the images c, b', b'' by first finding the image of O in AC.

These images all lie on a circle of which A is the centre and the radius AO. For the two right triangles OAB and bAB, having the two sides and the included angle of the one equal to the two sides and the included angle of the other, are equal. Therefore bA is equal to OA. In the same way it may be shown that bA equals $c'A$, $c'A$ equals $b''A$, etc. All the images are therefore equidistant from A and lie on the circle of which A is the centre and OA the radius.

When the mirrors are parallel the radius is infinite, the number of images is theoretically infinite, and they are all situated on a straight line drawn through the object and perpendicular to the mirrors. Practically the number of images is limited by the rapid decrease in the intensity of the light.

If the angle θ between the two mirrors is an aliquot portion of four right angles, or $\dfrac{2\pi}{\theta} = n$, then when n is even the number of images, including the object, is n; when n is odd the number of images is $n + 1$. From the last article it will be seen that for every pair of reflections from the two mirrors the ray suffers a deviation of twice the angle between the mirrors; and when it has changed its course by 180° it passes out between the two mirrors without further reflection. But at each reflection an image is formed. Therefore the number of images for each series, starting first with one mirror and then the other, will be $2\dfrac{\pi}{2\theta}$, and the number for both series will be $\dfrac{2\pi}{\theta}$. If θ is contained into π an exact number of times, or $\dfrac{2\pi}{\theta}$ is an even number, the last two images of the two series then coincide, so that the entire number is n, including the object.

If $\dfrac{2\pi}{\theta}$ is odd, that is if $\dfrac{\pi}{\theta}$ is not an integer, then the last two images of the two series do not coincide and the entire number, inclusive of the object, is $n + 1$.

179. Path of a Ray from any Image to the Eye. — Suppose the eye to be at the point I (Fig. 110); it is required to find the path of a ray from the object to the eye for the third image b''. Draw a line connecting b'' and I.

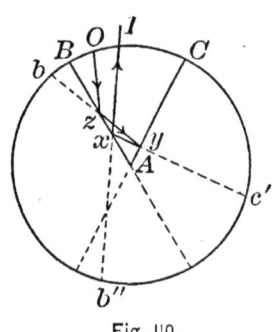

Fig. 110.

It intersects the mirror AB at x. Connect x and the next preceding image c' in the same series by the line intersecting the mirror AC in the point y. From y draw the line yb to the next preceding image, crossing the mirror AB at z. Finally join z and O. The path of the ray is $OzyxI$. If the image b'' is visible, the light enters the eye in the direction in which it is seen, viz., $b''I$. But the light traverses that path only up to the mirror at the point x, which is the point of last reflection. Now as b'' is the image of c', and c' acts in all respects as the true object for b'', the light must have reached the point x as if coming from c'. Therefore, the line xc' is drawn. But this intersects the other mirror in y, indicating the point where the next preceding reflection took place. The same process leads back finally to the object.

To find the path of a ray for any image from the object to the eye, draw a line from the eye to the image; from its intersection with the corresponding mirror, draw a line to the next preceding image of the same series; from the intersection of this line with the corresponding mirror, draw a line to the next preceding image, and so on, till a line is drawn from the last intersection to the object. The portions of the lines so drawn, lying in front of both mirrors, will be the path of the ray.

If the first line drawn does not intersect the mirror in which the given image is formed, then the eye is not in a position to view that image.

180. Deviation produced by the Rotation of a Plane Mirror. — If a plane mirror on which a ray of light falls be turned through an angle about an axis perpendicular to the plane of incidence, the reflected ray will be turned through twice that angle. Let a ray of light AM be incident normally on the mirror (Fig. 111); it will then retrace its path. If the mirror is now turned through the angle θ, the normal is turned through the same angle, so that the angles of incidence and reflection are now both equal to θ. The deviation is then the angle AMB, or 2θ.

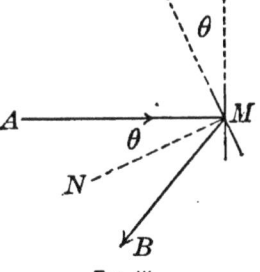

Fig 111.

On this principle a plane mirror is very extensively used to indicate, by the change in direction of the reflected ray, the motion of the movable system of the instrument to which the mirror is attached. The reflecting galvanometer, for the detection or measurement of minute currents of electricity, takes its name from the mirror which is attached to the needle, and which indicates the slightest rotational movement.

181. Concave Spherical Mirrors (L., 40; T., 64). — Reflection from each element of a curved surface takes place in accordance with the fundamental law of reflection. A beam of incident rays gives rise therefore to a system of reflected rays which can be geometrically determined.

Among curved mirrors spherical ones are the simplest and at the same time the most important.

They usually take the form of a spherical cap, polished either on the interior or the exterior. The former are called *concave;* the latter *convex.* The principal axis is the right line joining the centre of curvature and the *pole*, or central point of the spherical cap.

Let *AB* (Fig. 112) be a section of a concave spherical mirror through the principal axis *A U.* Let *U* be a luminous point. It is required to find the formula connecting its distance from the mirror with that of its image. If a ray

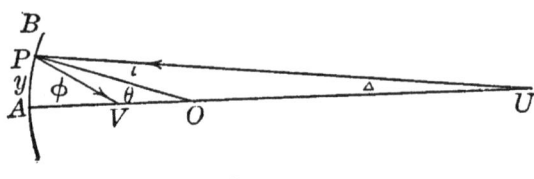

Fig. 112.

from *U* meet the mirror at *P* it will be reflected across the axis at *V*, so that the radius *OP*, which is the normal at the point of incidence, shall bisect the angle *UPV.*

Let the angle at *U* be represented by Δ, the angles of incidence and reflection by *i*, the acute angle at *O* by θ, and the acute angle at *V* by ϕ. Then because θ is external to the triangle *POU* and ϕ is external to the triangle *PVO*, we have

$$\theta = i + \Delta, \quad \cdots \cdots \quad (a)$$

$$\phi = i + \theta. \quad \cdots \cdots \quad (b)$$

Subtracting (*b*) from (*a*)

$$\theta - \phi = \Delta - \theta,$$

or
$$2\theta = \Delta + \phi. \quad \cdots \cdots \quad (c)$$

˜ If now P is very near A, the pole of the mirror, the angles θ, ϕ, and Δ are very small, and their tangents may be put equal to the angles themselves. Let AU be represented by p, AV by p', and the radius PO by r. Also let y equal PA. Then from (c)

$$\frac{2y}{r} = \frac{y}{p} + \frac{y}{p'}.$$

Whence

$$\frac{1}{p} + \frac{1}{p'} = \frac{2}{r}.$$

Since y does not appear in this equation, it follows that the distance of V from the mirror, corresponding to a given distance of U, is independent of the position of the point P, to the extent of the approximation made that the tangent of an angle is equal to the angle itself. The physical interpretation of this fact is that *for small angles of incidence* all of the rays from U, incident upon the mirror, are reflected so as to pass through the common point V. Hence V is the focus of the radiant point U, and the two are called *conjugate foci*. V is a *real image* because the rays actually pass through it and may be received upon a screen. The relation between U and V is conjugate because if V were the radiant point the focus after reflection would be U.

182. Principal Focus and Discussion of the Formula. — (a) Since the sum of the reciprocals of p and p' is a constant $\frac{2}{r}$, it follows that as p increases p' decreases, and when p becomes infinite p' equals $\frac{r}{2}$. Hence parallel rays from an infinitely distant source come to a focus at a point midway between the centre of curvature and the mirror. The focus for *parallel* incident rays is called the *principal focus*.

(*b*) When *p* decreases *p′* increases, or the object and image approach each other. When *p* equals *p′*, $\frac{2}{p} = \frac{2}{r}$, or object and image coincide at the centre of curvature.

(*c*) When *p* is less than *r* and greater than $\frac{r}{2}$, *p′* is greater than *r*, or object and image have exchanged places.

(*d*) When *p* is less than $\frac{r}{2}$, $\frac{1}{p}$ is greater than $\frac{2}{r}$, and *p′* is therefore negative. The image is then behind the mirror and hence virtual. Distances in front of the mirror are considered positive and those at the back negative. When *p* is at the principal focus the image is at an infinite distance in either direction. As *p* approaches the mirror *p′* also approaches it from behind, and the two again meet at the surface of the mirror.

183. Formation of Images in a Concave Mirror. — Fig. 113 is the graphical construction for the image when the object *AB* is beyond the centre of curvature *C*.

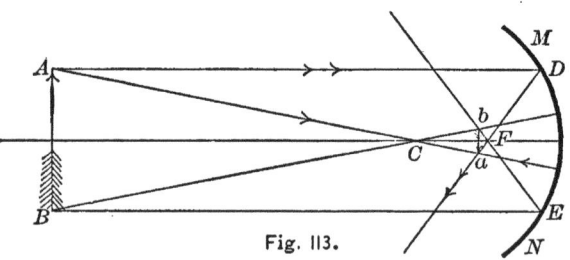

Fig. 113.

For the purpose of finding the image of any point of the object it is necessary to trace the path of two rays only, and to find their intersection after reflection. For convenience the two rays selected are parallel to the principal axis, and along a secondary axis, respectively. A secondary axis is any right line passing through the centre of curvature. The ray *AD* after reflection passes through the principal focus F; the ray AC is reflected directly back on its own path. The intersection of these two reflected rays is at a, and a is therefore the focus conjugate to A. In the same way the two rays BE and BC intersect after reflection at b, the conjugate focus or image of B. Therefore ab is the image of AB. It is inverted and real.

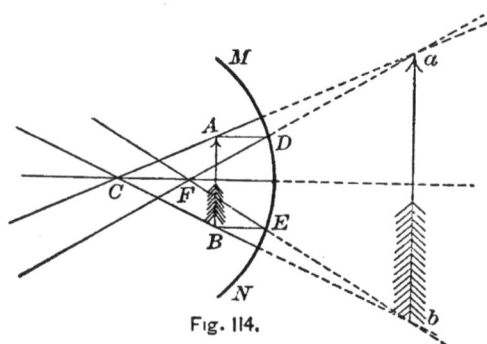

Fig. 114.

Fig. 114 is the construction for a virtual image in a concave mirror. The ray AD parallel to the principal axis is reflected through the principal focus F. The ray CA retraces its path after reflection. The two rays themselves do not meet, but if the lines representing their paths are produced behind the mirror they meet at a. This is therefore the virtual image of A. Similarly b is the virtual image of B, and ab is the image of AB. It is erect, enlarged, and virtual.

184. Caustics by Reflection (T., 70; P., 90; V., II, 376). — When the angular opening of the mirror is large the approximation made in the last article is no longer admissible. Parallel rays do not all meet at the principal focus after reflection, and a luminous point no longer has

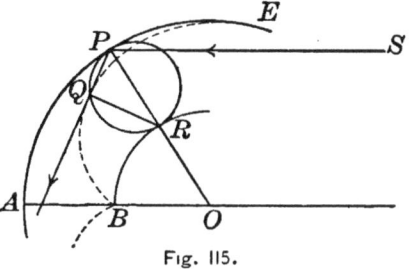

Fig. 115.

a definite image. The intersections of the reflected rays then form a luminous surface called a *caustic*. A section of this curve is familiar to every one who has looked at a cup of milk illuminated by a bright light.

The simplest case of a caustic by reflection is furnished by parallel rays of light falling on a concave spherical mirror. Let AP (Fig. 115) be a section of the mirror through its centre O, and let SP be one of a system of rays parallel to AO.

Then since all the rays are symmetrical about AO, if we find the section of the caustic in the plane of the figure, a caustic surface will be generated by revolving this curve about AO as an axis.

Join P and O and let PQ be the path of the reflected ray. Bisect PO in R and on PR as a diameter draw a circle; also with O as a centre and with radius OR construct another circle RB. The two circles touch at R. The angle QPR equals the angle ROB. But the arc QR subtends QPR at the *circumference*, and the arc BR subtends ROB at the *centre* of a circle of double radius. Hence the arcs QR and RB are equal; and if the small circle should roll on the inner one the point Q would ultimately coincide with B, and would describe the epicycloid indicated by the dotted curve. Moreover, the point of contact R at any instant is fixed, and Q is therefore moving at right-angles to QR or in the direction of the reflected ray PQ. Hence all the reflected rays touch the epicycloid; and since all the reflected rays are tangent to the required caustic, the epicycloid is therefore a section of the caustic surface; for the reflected rays cross everywhere on this section. At B is a cusp which is the principal focus of the mirror. Not all parallel rays after reflection pass through this focus. The effect of this inexactness upon the image is known as *spherical aberration*.

The caustic is tangent to the principal axis at *B* and to the mirror at *E*.

This example furnishes the data for the explanation of the two *focal lines*, due to a small pencil of rays incident at some point *P ;* this pencil has no focus. Every reflected ray *PQ* passes through the axis *OA* of the mirror. But *Q* is the intersection of two successive rays in the plane of the figure. When the caustic is made to rotate about *AO* the point *Q* describes a circle at right angles to the plane of the paper. Hence a small pencil of rays incident about *P* is reflected so as to pass through a short line at *Q*, perpendicular to the plane of the paper, and through a short line along the axis *AO*, where *PQ* intersects it. Between these two mutually perpendicular focal lines is the *circle of least confusion*, which is the section of the pencil in which the rays are most closely crowded together.

On account of the spherical aberration of concave spherical mirrors, they cannot be used for astronomical purposes; but if they have a parabolic section, all rays falling on them parallel to the axis will be reflected so as to pass exactly through the focus.

185. Convex Spherical Mirrors. — The formula of Art. 181 is applicable to a convex mirror. In this case the centre of curvature and the radiant point are on opposite sides of the mirror. If distances in front of the mirror are still considered positive, *r* for this case is negative. The formula then becomes

$$\frac{1}{p} + \frac{1}{p'} = -\frac{2}{r}.$$

All the images of a point in a convex mirror are virtual; for, while *p* is positive, *p'* must be negative, since the sum of the reciprocals of *p* and *p'* is a negative quantity.

In Fig. 116 the polished surface is on the convex side,

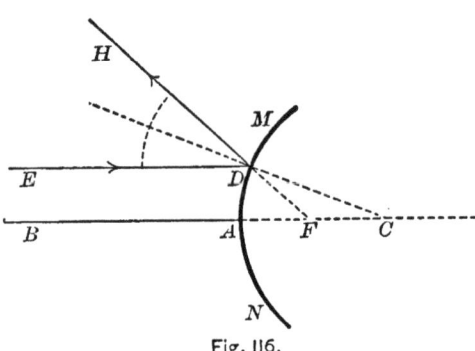

Fig. 116.

C is its centre of curvature, *ED* is a ray parallel to the principal axis, and *DH* is its path after reflection. It comes as if from *F*, which is therefore the principal focus midway between *A* and *C*.

Fig. 117 shows the construction for the image in a convex mirror. *AD* is a ray parallel to the principal axis, whose direction after reflection passes through the principal focus. *AC* is a ray along the secondary axis through *A*. These two lines meet in *a*, which is then the v i r t u a l image of *A*. In the same way the virtual image of *B* may be found at *b*. The image is erect, smaller than the object, and virtual.

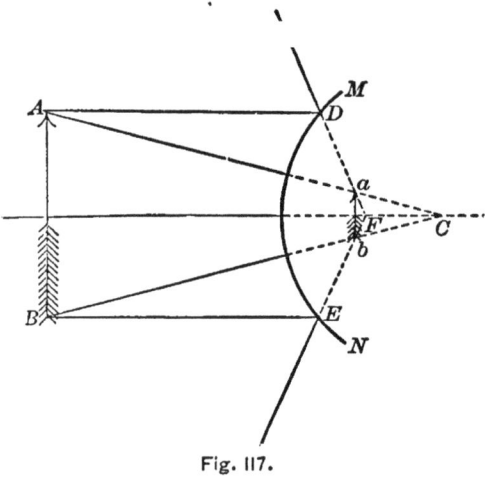

Fig. 117.

PROBLEMS.

1. An object 6 cms. long is placed symmetrically on the axis of a convex spherical mirror at a distance of 12 cms. from it; the image formed is 2 cms. long. What is the focal length of the mirror?

2. A candle flame is placed at a distance of 30 cms. from a concave mirror made from a sphere of 30 cms. diameter. Find the position of the image. Is it erect or inverted?

3. The radius of a convex mirror is 20 cms. If the linear dimensions of an object be twice those of the image, where must each be situated?

186. Refraction (P., 73; V., II, 387; L., 56). — When a luminous ray passes from one transparent medium into another it undergoes in general a change in direction at the surface of separation of the two media. The portion entering the second medium is said to be *refracted*.

Let *MN* (Fig. 118) be the surface of separation, *BA* the incident, and *AC* the refracted ray. The plane of incidence is the plane *BAF* containing the incident ray and the normal to *MN* at the point of incidence. *CAG* is the plane of refraction.

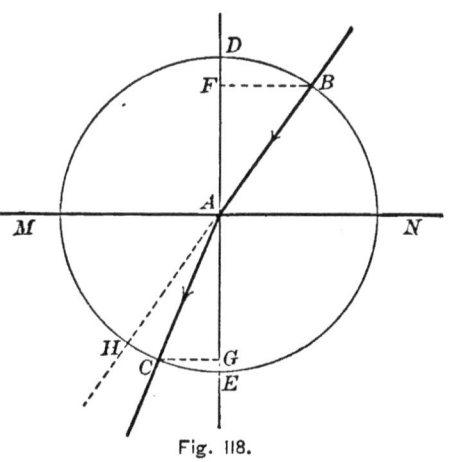

Fig. 118.

If the new medium into which the light is propagated is isotropic, *the plane of incidence and the plane of refraction coincide, and the ratio of the sines of the angles of incidence and refraction is a constant.* If the second medium is optically denser than the first, the refracted ray is deflected toward the normal, as in the figure. If the radius of the circle is unity, *BF* and *CG* represent the sines of the angles of incidence and refraction respectively. Denoting the angles of incidence and refraction by *i* and *r* respectively, the law of sines in single refraction is

$$\frac{\sin i}{\sin r} = \mu.$$

The constant ratio μ is called the *index of refraction.*
When light passes from a vacuum into any medium, this
ratio is called the *absolute index* of refraction; but when
it passes from one medium into another, it is called the
relative index. The relative index from medium a to
medium b is equal to the absolute index of b divided by
that of a.

**187. Law of Refraction deduced from the Undu-
latory Theory (T., 175; P., 74; B., 426).** — Let AB
(Fig. 119) be the trace of the
incident plane wave, and AC
that of the surface of separation, both planes being perpendicular to the plane of the
paper. Let v be the speed of
light in the first medium and v'
that in the second. Then if t is
the time required for light to

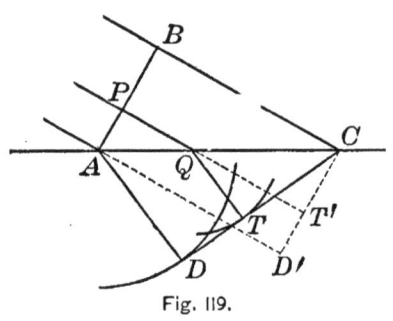

Fig. 119.

traverse the distance BC, vt equals BC equals AD'. If
there had been no change of medium the wave-front at the
end of the time t would have had the position $D'C$, parallel
to AB.

But the disturbance in the second medium travels with
a speed v'. Therefore a sphere drawn with a radius $AD =$
$v't$ will limit the spread of the wave from A in the second
medium. In the same way the disturbance from the point
P will travel to Q, which then becomes a new centre of
disturbance, and this will extend into the second medium at
the end of the time t a distance QT, such that $\dfrac{QT}{QT'} = \dfrac{v'}{v}$. All
circles representing the traces of such spherical waves
ultimately intersect along CD, which is the trace of a plane

drawn through C tangent to the first circle with radius AD. CD is therefore the trace of the new wave-surface in the second medium.

The angles of incidence and refraction are BAC and ACD respectively. Hence

$$\mu = \frac{\sin i}{\sin r} = \frac{\sin BAC}{\sin ACD} = \frac{BC}{AD} = \frac{vt}{v't} = \frac{v}{v'}.$$

The index of refraction is therefore equal to the ratio of the speed of light in the first medium to its speed in the second. The undulatory theory thus gives a satisfactory explanation of the law of single refraction. If the speed of light in a vacuum be taken as the unit, the absolute index of refraction for any medium will be the reciprocal of the speed in that medium.

Michelson obtained for the ratio of the speeds in air and water the value 1.33, and for air and carbon disulphide 1.758. These values are a close approximation to the relative indices of refraction in the two cases.

If the speed of light in medium a is v_a and in medium b is v_b, then the absolute index for a is $\frac{1}{v_a}$ and for b, $\frac{1}{v_b}$. But the relative index from a to b is $\frac{v_a}{v_b}$; and this is equal to the absolute index of b divided by that of a, or $\frac{\mu_b}{\mu_a}$.

188. Newtonian Explanation of Refraction. — The Newtonian theory of light ascribes the change in the direction which a ray undergoes at the surface of separation between two media to the greater attraction of the denser medium for the corpuscles of light. The resultant of all this attraction on a corpuscle as it approaches this surface must be along a normal; therefore the component of the motion parallel to the surface of the new medium will be

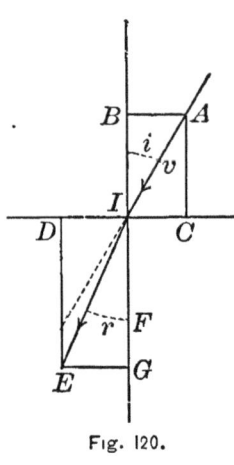

Fig. 120.

unaffected. Let v and v' be the speed of light in the two media, denoted by AI and IE in Fig. 120, and let i and r be the angles of incidence and refraction. Then the component of v, parallel to the surface DC, will be $v \sin i$, and of v' it will be $v' \sin r$. The normal component of v or BI, equal to IF, will be increased by some quantity FG, due to the attraction of the denser matter for the corpuscle. Therefore, completing the parallelogram DG, its diagonal IE represents the speed of light in the new medium. It is greater than v for the rarer medium.

Placing the two expressions for the tangential component equal to each other,

$$v \sin i = v' \sin r.$$

Whence

$$\mu = \frac{\sin i}{\sin r} = \frac{v'}{v}.$$

But by the undulatory theory the index of refraction is $\dfrac{v}{v'}$, and v is greater than v'. The emission theory, therefore, requires that light travel faster in the denser medium, like water or glass, than in the rarer; the undulatory theory leads to the opposite conclusion. Foucault's experiment showed conclusively that the emission theory is untenable, for light travels slower in water than in air.

189. Refraction at a Plane Surface (T., 85; V., II, 400). — Let MO (Fig. 121) be the normal to the plane

surface, and let IB be an emergent ray from the luminous point O. Project the emergent ray downwards till it intersects the normal at I. Then BIA is the angle of incidence (the angle in the less dense medium) and BOA the angle of refraction.
Hence

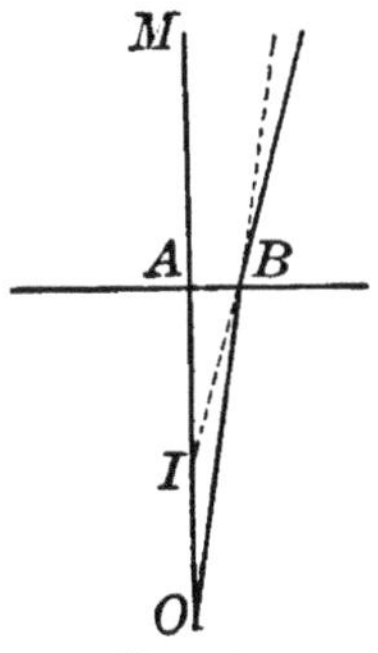

$$\mu = \frac{\sin BIA}{\sin BOA}.$$

But $\sin BIA = \sin BIO$, since these angles are supplementary; and as the sines of the angles of a triangle are proportional to the sides opposite,

$$\mu = \frac{BO}{BI}.$$

Fig. 121.

When BO is nearly normal, and the pencil of rays is only slightly divergent, BO is ultimately equal to AO, and BI to AI. But I is the position of the image of O. Looking along a normal line an object at O will therefore appear to be at I. For water μ is about $\frac{4}{3}$, and for glass $\frac{3}{2}$. An object in water cannot appear lower than $\frac{3}{4}$ its real depth, and one in glass not further than $\frac{2}{3}$ of its real distance from the surface. To an eye under water an object in air appears $\frac{4}{3}$ of its real distance from the surface.

Viewed obliquely the depth of water appears still less than $\frac{3}{4}$ of the actual depth. Hence the shoaling appearance of still water in which the bottom is visible. The images of an object under water will lie on a caustic surface with the cusp at $\frac{3}{4}$ the real distance of the object beneath the surface. The cusp will be on a line drawn normal to the surface and through the object.

PROBLEMS.

1. A straight rod is partially immersed in water. The image, viewed normally, is inclined 45° to the surface. If the index of refraction is $\frac{4}{3}$, what is the inclination of the stick?

2. If the angle of incidence is 60°, and the index of refraction is $\sqrt{3}$, find the angle of refraction.

3. If the index of refraction from air to water is $\frac{4}{3}$ and from air to glass $\frac{3}{2}$, what is the relative index from water to glass?

190. Construction for the Refracted Ray. — Let *MN* be the surface separating the two media, as air and water

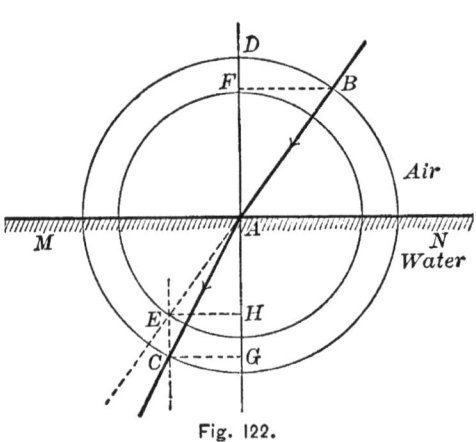

Fig. 122.

(Fig. 122). Let *BA* be incident on the second medium at *A*. With *A* as a centre draw two concentric circles with radii proportional to the speed of light in the two media. Their ratio is then the index of refraction. Produce the line *BA* till it intersects the inner circle in *E*, and through *E* draw *EC* parallel to the normal *DA*. It intersects the outer circle in *C*. Draw *AC;* it is the path of the refracted ray.

The triangles *EAH* and *BAF* are similar. Therefore

$$\frac{BF}{CG} = \frac{BF}{EH} = \frac{AB}{AE} = \mu.$$

But *BF* and *CG* are proportional to the sines of the angles *BAF* and *CAG*, and the angle *BAF* is the angle of incidence. Therefore the angle *CAG* is the angle of refraction.

\ When the ray passes into a less dense medium, through the intersection of the ray produced with the *larger* circle draw a parallel to the normal, cutting the smaller circle. The line through this intersection and the point of incidence is the refracted ray.

191. The Critical Angle (T., 86; P., 76). — When light passes from one medium into another of smaller optical density, as from water into air, the ray is refracted away from the normal, and the angle in the first medium is less than in the second. When the angle in the second medium becomes a right angle and the ray just grazes the surface, the angle in the first medium is a maximum and is called the *critical angle.*

Thus the ray RO in water (Fig. 123) emerges in air in the direction OS; for the ray LO the corresponding ray in the rarer medium is along the surface OB. If the incident ray in the denser medium makes a greater angle with the normal than LON' it can no longer emerge into the second medium, but undergoes *total internal reflection.* Thus the ray

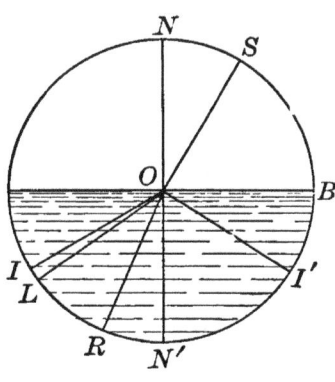

Fig. 123.

IO is totally reflected in the direction OI'. LON' is the critical angle. For a smaller incident angle part of the light is reflected and part refracted; for larger angles of incidence it is all reflected.

To determine the critical angle for any medium whose relative index is μ, we have

$$\mu = \frac{\sin 90°}{\sin x}.$$

Whence
$$\sin x = \frac{1}{\mu},$$

or *the sine of the critical angle is the reciprocal of the index of refraction.*

For water the critical angle is . . . . . 48° 27′ 40″
For crown glass about . . . . 41° 10′
For chromate of lead . . . . 19° 49′

The total internal reflection of light is beautifully shown by focusing a beam of light by means of a lens *L* (Fig.

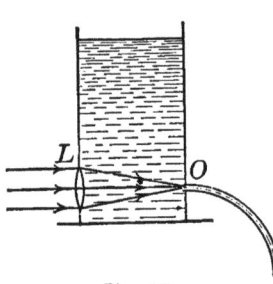

Fig. 124.

124) on the interior of a smooth jet at the point of issue *O* from the side of the vessel. The angle of incidence on the interior surface of the jet then exceeds the critical angle, and the beam is reflected from side to side along the stream like sound in a speaking-tube. If colored glass is interposed the stream is colored and presents a beautiful appearance. It becomes visible by means of the diffused light irregularly reflected from fine matter suspended in the water.

PROBLEMS.

1. In Iceland spar there are two refracted rays (225). The indices of refraction for the two are 1.658 and 1.486. Find the critical angle for each.

2. For chromate of lead the least index of refraction is 2.5; the greatest, 2.95. To which index does the above critical angle correspond? What is the other critical angle?

3. What is the greatest apparent zenith distance which a star can have as seen by an eye under water?

192. **General Construction for Refraction at a Single Surface (D., 125).** — Let AB (Fig. 125) be an advancing wave-front and CD the bounding surface of the denser medium. Through a number of points of AB draw normals AE, BF, etc. Make all these normals equal to one

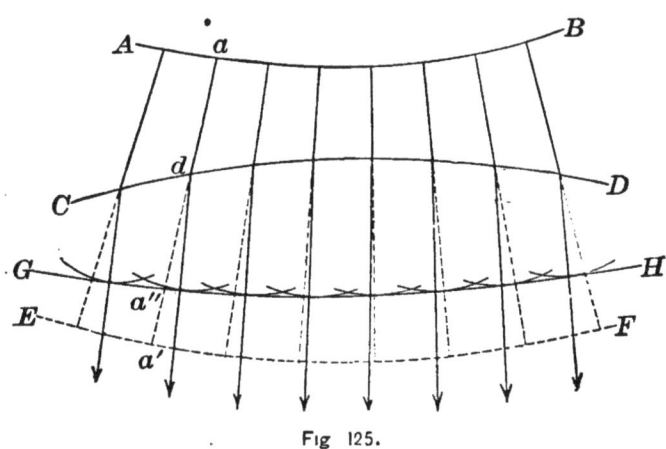

Fig 125.

another and connect their extremities by a plane curve. If t is the time required to traverse these normals with the speed v in the first medium, then this plane curve $Ea'F$ would have been the wave-front at the end of the interval t if no denser medium had intervened.

To find the wave-front in the second medium, take any point d where the normal from a intersects the bounding surface of the denser medium as a centre, and with a radius da'', which bears to da' the same ratio that v', the speed in the second medium, bears to v, the speed in the first, draw an arc of a circle. This defines the limit to which the disturbance from a has penetrated into the denser medium at the end of the time t. Draw circles in the same way from all the other intersections of the normals with the bounding surface. The common tangential curve, enveloping all the circles representing the secondary waves, will be the wave-front in the new medium.

From the centres of the secondary waves normals may be drawn to this new wave-front. They will not in general coincide with *aa'* and its fellows, but will show by their deviation from them the amount of refraction at each point of the surface.

193. Construction for the Refraction of a Spherical Wave at a Plane Surface (D., 126; P., 83). — Let a spherical wave from a centre *O* (Fig. 126) fall upon the plane surface *AB* of a transparent medium. After a short interval *t* with the same medium this wave would have taken the position *ADB*, a sphere of radius *OD* and centre *O*. But the speed in the second medium is reduced along each normal in the ratio of *v'* to *v* or $\frac{1}{\mu}$.

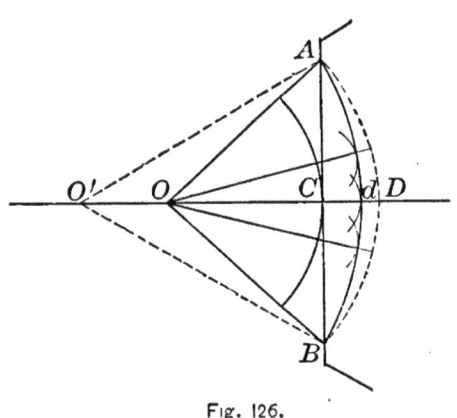

Fig. 126.

Therefore with *C* as a centre and a radius *Cd* such that $CD = \mu \cdot Cd$, describe a small circle. In the same way any other point of the surface *AB* will be the centre of a corresponding secondary wave. The refracted wave will be the envelope of all these secondary spherical waves, and the new wave-front will be *AdB*. The incident wave is therefore flattened down into another of smaller curvature. If the width *AB* be very small compared with *OC*, the new wave-front will be approximately spherical, and the arc *AdB* will be an approximate circle with its centre at *O'*.

If r and r' denote the radii, OD and $O'd$, of the incident and refracted waves

$$\overline{AC}^2 = 2r \cdot CD = 2r' \cdot Cd \qquad \text{nearly.}$$

But $$CD = \mu \cdot Cd;$$

therefore $$r' = \mu r,$$

or the radius of the refracted wave is approximately μ times that of the incident wave. The true wave-surface in the second medium is hyperbolic.

194. Refraction through a Prism (T., 96; P., 80; D., 490). — A prism for optical purposes consists of a transparent medium bounded by two planes enclosing an angle which is less than twice the critical angle for the substance. This angle A (Fig. 127) is called the *refracting angle* of the prism.

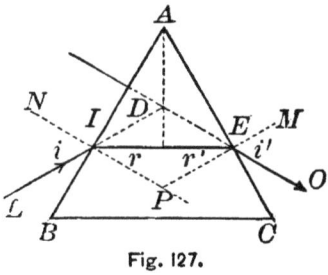

Fig. 127.

Since light is bent toward the normal on entering a medium of higher refrangibility, and away from it when passing into one of lower refrangibility, the path of a ray of homogeneous light through the prism may be such as *LIEO*. I is the point of incidence and E of emergence.

The precise path of the ray through the prism and after emergence may be found, when the index of refraction is known, by the method of Art. 190. This is shown applied to a prism in Fig. 128. Two arcs are drawn with the common centre E, and radii in the

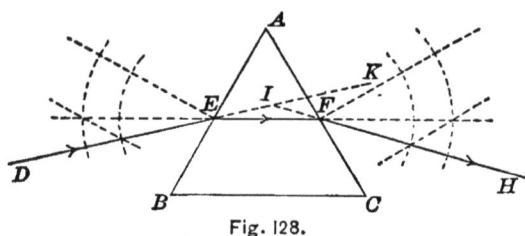

Fig. 128.

ratio of 1 to μ. Through the intersection of the incident ray with the inner arc draw the dotted line parallel to the normal through E. Through its intersection with the outer arc draw a line to E and produce through the prism. This is EF, the path through the prism. In a similar way the two arcs about F serve tò draw the emer- gent ray. It will be observed that the construction line drawn parallel to any normal must intersect the arc of smaller radius at the same point as the ray in the rarer medium, and the arc of larger radius at the same point as the ray in the denser medium.

The deviation at the first surface of the prism (Fig. 127) is $i - r$; at the second surface, $i' - r'$. Therefore the total deviation is

$$D = i - r + i' - r' = i + i' - (r + r').$$

The angle of the prism A equals the angle between the normals ; and since this is the angle external to the triangle IPE at P, it equals the sum of the two interior opposite angles r and r'. Therefore

$$A = r + r'.$$

If now the path of the ray through the prism be symmetrical with respect to the two faces of the prism, which is the condition of minimum deviation (195), then $i = i'$ and $r = r'$. Therefore $A = 2r$ and

$$D = 2i - 2r = 2i - A.$$

Hence
$$i = \frac{A + D}{2},$$

and
$$\mu = \frac{\sin i}{\sin r} = \frac{\sin \frac{1}{2}(A + D)}{\sin \frac{1}{2} A}.$$

This is the formula commonly used for measuring the index of refraction for any ray whose *minimum* deviation is D.

When the angle of the prism is very small an approximate formula may be found by making the angles A and D equal to their sines. Then

$$\mu = \frac{A+D}{A} = 1 + \frac{D}{A},$$

and $D = A(\mu - 1).$

If a perpendicular be dropped from A upon IE when $r = r'$, each half of the angle A is equal to the angle of refraction r. If A were equal to or greater than twice the critical angle for the transparent substance, r would be equal to or greater than this critical angle, and the ray would suffer total internal reflection instead of refraction. Therefore the angle of the prism must be less than double the critical angle.

For crown glass the critical angle is about 41° 10'. Hence if a prism of crown glass have a section perpendicular to its refracting edge of a right-angled isosceles triangle BAC (Fig. 129) the refracting angle will be more than twice the critical angle, and a ray DE incident normally on either face adjacent to the right angle will have an internal angle of incidence greater than the critical angle, and will be totally reflected at E. For flint glass the critical angle is still smaller.

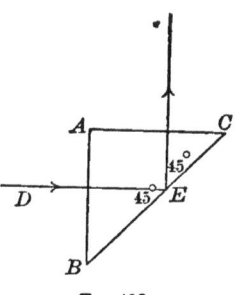

Fig. 129.

195. Construction for Deviation (A. and B., 408; Deschanel, 1008).

— The following geometrical construction furnishes a simple method of determining the deviation for any angle of incidence:

(*a*) Describe two circular arcs about a common centre

O (Fig. 130), the ratio of their radii being the index of refraction. Draw *OA* for the direction

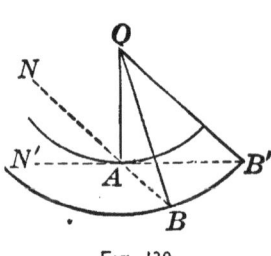

Fig. 130.

of the incident ray and through its intersection with the inner arc draw *NA* for the normal. Then the angle *NAO* is the angle of incidence *i*. Produce *NA* till it intersects the outer arc in *B* and connect *B* and *O*. Then *OBN* is the angle of refraction and the deviation is *AOB*. For

$$\frac{\sin OAN}{\sin OBN} = \frac{\sin i}{\sin OBN} = \frac{OB}{OA} = \mu.$$

Therefore *OBN* is the angle of refraction, and *AOB* is the difference between *OAN* and *OBN*, or the deviation.

Draw *N'B'* making an angle of 90° with *OA*, and join *B'* and *O*. Then *OB'N'* is the critical angle, for it is the angle of refraction corresponding to an angle of incidence of 90°.

(*b*) To determine the deviation for refraction at both surfaces of a prism, draw the circular arcs, the incident ray, and the normal at the first surface as before. Then through *B* draw *BN'* (Fig. 131) to represent the normal at the second surface. Then *OB* is the *direction* (not the path) of the ray through the prism, and *OA'* is the direction of the emergent ray. Since *AOB* is the deviation at the first surface,

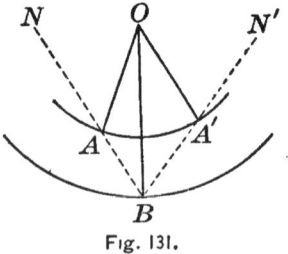

Fig. 131.

A'OB in the same way is the deviation at the second surface, and the total deviation is the angle *AOA'*.

(*c*) The angle *NBN'*, the angle between the normals, is equal to the refracting angle of the prism. Further

the angle AOA' is measured by the circular arc AA'. To
find the conditions under which this shall be a minimum,
we must ascertain what position of the constant angle
ABA' will give the shortest arc

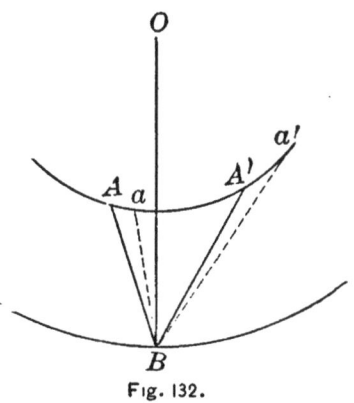

Fig. 132.

AA'. Let ABA' and aBa' (Fig.
132) be two consecutive positions, BA' and Ba' being greater
than BA and Ba. Then the
arc aa' is greater than AA', both
because A' is farther from B
than A and because $A'a'$ cuts
across the angle more obliquely
than Aa; and $A'a' - Aa$ is the
increase in the length of the arc
which measures the deviation. The deviation is therefore
increased by changing the position in such a way as to
make BA and BA' depart further from equality. It is
then a minimum when BA and BA' are equal, or when
the incident and emergent rays make equal angles with
the normals at the two surfaces of the prism.

**196. Refraction at Spherical Surfaces (T., 107; P.,
84).** — Let O (Fig. 133) be the centre of curvature of the
spherical surface
AB of the denser
medium. Let U
be the p o i n t-
source of homo-
geneous light,
that is, light of

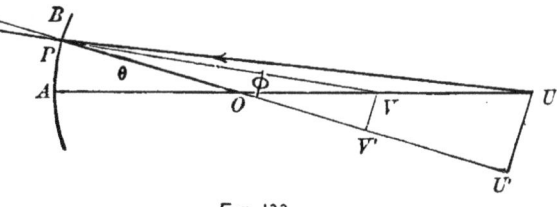

Fig. 133.

one color, whose index of refraction is μ. Let UP be the
incident ray. Produce it backward after refraction to
meet the principal axis in V.

Then since UPO is the angle of incidence and VPO the angle of refraction,

$$\frac{\sin i}{\sin r} = \frac{\sin UPO}{\sin VPO} = \frac{UU'}{PU} \div \frac{VV'}{PV} = \frac{UU'}{VV'} \cdot \frac{PV}{PU} = \mu.$$

But from the figure $\quad \dfrac{UU'}{VV'} = \dfrac{OU}{OV}.\quad$ Substituting,

$$\mu = \frac{OU}{OV} \cdot \frac{PV}{PU} = \frac{OU}{OV} \cdot \frac{AV}{AU}, \text{ when } P \text{ is very near } A.$$

If we use the same notation as in Art. 181, putting p for AU, p' for AV, and r for AO, then

$$\mu = \frac{p-r}{p'-r} \cdot \frac{p'}{p}.$$

Reducing, $\qquad \dfrac{\mu}{p'} - \dfrac{1}{p} = \dfrac{\mu-1}{r}. \quad \cdots \cdots \quad (a)$

If $\mu = -1$, this becomes the formula for the concave spherical mirror. For reflection μ must have the value unity, and the minus sign expresses the reversal of the ray in reflection.

If the refracted ray meet another spherical surface, with its centre of curvature on the line OA, it will undergo another refraction. If the second surface be very near the first one, the distance of the virtual point-source V from it will be p'; and if we put q for the distance of the conjugate focus from the second surface, and r' for the radius of curvature of this surface, we have, since the refraction is from dense to rare, and the index of refraction is therefore $\dfrac{1}{\mu}$,

$$\frac{\frac{1}{\mu}}{q} - \frac{1}{p'} = \frac{\frac{1}{\mu} - 1}{r'},$$

or $\qquad \dfrac{1}{q} - \dfrac{\mu}{p'} = \dfrac{1-\mu}{r'}. \quad \cdots \cdots \quad (b)$

Adding equations (*a*) and (*b*) and

$$\frac{1}{q} - \frac{1}{p} = (\mu - 1)\left(\frac{1}{r} - \frac{1}{r'}\right).$$

This is the approximate formula for a very thin *lens*, a transparent body bounded by two curved surfaces, or one curved and one plane. The distances *p* and *q* are the *conjugate focal distances*.

197. The Principal Focus. — If the source *U* is at an infinite distance, *p* is infinity, and the formula becomes

$$\frac{1}{q} = (\mu - 1)\left(\frac{1}{r} - \frac{1}{r'}\right).$$

If, for this particular case, we put *f* for the distance *q*, then

$$\frac{1}{f} = (\mu - 1)\left(\frac{1}{r} - \frac{1}{r'}\right).$$

The distance *f* is called the *focal length*. The rays coming from an infinite distance are parallel and they converge after refraction to the *principal focus*.

If the curvature of the two faces is the same, and *r* is negative,

$$-\frac{1}{f} = (\mu - 1)\frac{2}{r},$$

for *double convex lenses*.

If *r'* is negative,

$$\frac{1}{f} = (\mu - 1)\frac{2}{r},$$

for *double concave lenses*.

198. The Sign of the Quantity *f*. — In the sense in which we have used the quantity *μ* it is always greater than unity. Therefore the formula shows that *f* follows the sign of $\left(\dfrac{1}{r} - \dfrac{1}{r'}\right)$, the difference of the curvatures of the two surfaces. This focal distance is positive for all lenses thinner at the centre than at the edges, or the principal focus lies on the same side of the lens as the source of light; such lenses produce a divergence of the beam.

Distances measured from the lens opposite to the direction in which the light traverses it are regarded as positive. Consider the three forms of Fig. 134. In **1**, r is

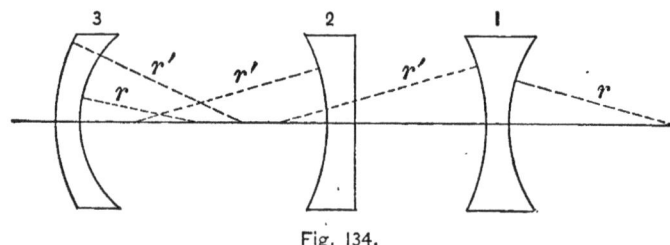

Fig. 134.

positive and r' negative. Therefore $\dfrac{1}{r} - \dfrac{1}{r'}$ becomes $\dfrac{1}{r} + \dfrac{1}{r'}$, a positive quantity. In **2**, the radius r' is negative and r is infinite. Therefore $\dfrac{1}{r} - \dfrac{1}{r'}$ is positive. In **3**, both r and r' are positive, but r is smaller than r'; therefore $\dfrac{1}{r} > \dfrac{1}{r'}$ and $\dfrac{1}{r} - \dfrac{1}{r'}$ is again positive. All these lenses are therefore *diverging* because the focal distance is positive.

For all lenses thicker at the centre than at the edges $\dfrac{1}{r} - \dfrac{1}{r'}$ is negative, or the principal focus and the source are on opposite sides of the lens.

Consider the three forms of Fig. 135. In **1**, r is negative and r' is positive. Therefore $\dfrac{1}{r} - \dfrac{1}{r'} = -\dfrac{1}{r} - \dfrac{1}{r'}$, a negative quantity. In 2, either r is negative and r' is infinite, or r is infinite and r' is positive, according to the direction

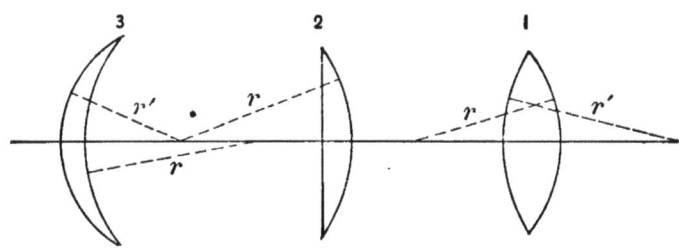

Fig. 13S.

toward which the curved side of the lens is turned; but in both cases $\frac{1}{r} - \frac{1}{r'}$ is negative. In 3, r is positive and greater than r', which is also positive. Hence $\frac{1}{r'} > \frac{1}{r}$ and $\frac{1}{r} - \frac{1}{r'}$ is negative. For all three forms the principal focus and the source are on opposite sides of the lens. They are then all *converging* lenses.

199. Images in Lenses. — Since the quantity $(\mu - 1)$ $\left(\frac{1}{r} - \frac{1}{r'}\right)$ is equal both to $\frac{1}{q} - \frac{1}{p}$ and to $\frac{1}{f}$, these latter quantities may be equated, giving

$$\frac{1}{q} - \frac{1}{p} = \frac{1}{f}.$$

In the last article we have seen that for diverging lenses f is always positive. Now since p, the distance of the object, is necessarily positive, q must be positive and smaller than p. The image is therefore on the same side of the lens as the object, it is virtual and nearer the lens than the object. Thus in Fig. 136, A is the source, and the ray AB, after passing through the lens, diverges from the axis as if it came from A'. A and A' are on the same side of the lens, and A' is nearer the lens than A.

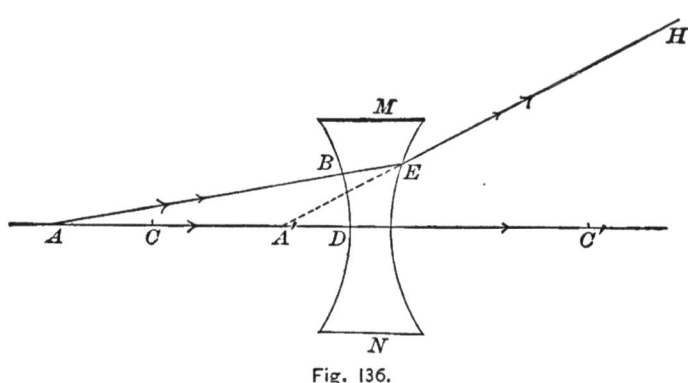

Fig. 136.

For converging lenses f has been shown to be negative. Then

$$\frac{1}{q} - \frac{1}{p} = -\frac{1}{f}.$$

When $\frac{1}{p} < \frac{1}{f}$, or where $p > f$, q must be negative, or object and image are on opposite sides of the lens; but when $\frac{1}{p} > \frac{1}{f}$, or when $p < f$, q is positive, or the image is on the same side of the lens with the object and is therefore virtual. The first case is illustrated by Fig. 137, in

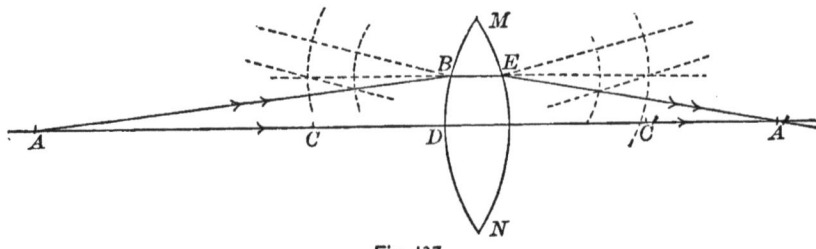

Fig. 137.

which the conjugate foci A and A' are on opposite sides of the double convex lens. Fig. 138 illustrates the second case. A and A' are both on the same side of the lens.

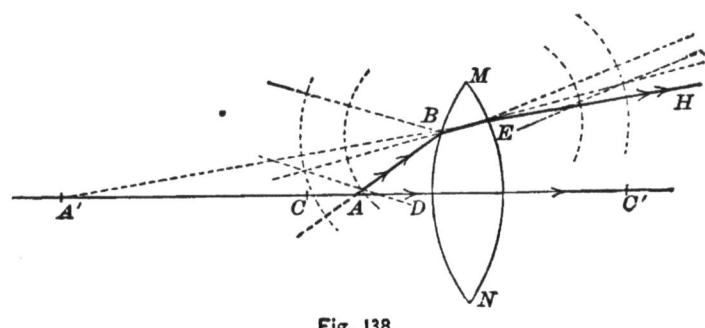

Fig. 138.

For real images q is negative. The formula for converging lenses may then be written after changing all the signs,

$$\frac{1}{q} + \frac{1}{p} = \frac{1}{f}.$$

When $p = q$ we have

$$\frac{2}{p} = \frac{1}{f}, \text{ or } p = 2f.$$

Object and image are then equidistant from the lens, and the distance between them is $4f$. They cannot approach nearer than this for a real image.

200. Image and Object at a Fixed Distance. — When the object, such as an incandescent lamp filament, and the screen on which the image is received are at a fixed distance, which must be greater than four times the focal length, there are two positions of the lens which will give an image. In the first the lens is nearer the object and the image is enlarged; in the second the two conjugate focal distances are exchanged, and the lens is nearer the image, which is then smaller than the object.

Let l be the distance between the object and the screen, and let a be the distance moved over by the lens in changing the focus from one image to the other.

Then $\qquad\qquad q + p = l.$

Also $\qquad\qquad q - p = a.$

Adding these equations and we have

$$q = \frac{l + a}{2}.$$

Subtracting $\qquad\qquad p = \frac{l - a}{2}.$

Therefore $\qquad \dfrac{1}{f} = \dfrac{2}{l + a} + \dfrac{2}{l - a} = \dfrac{4l}{l^2 - a^2},$

and hence $\qquad\qquad f = \dfrac{l^2 - a^2}{4l}.$

This formula furnishes a very satisfactory method of measuring the focal length of a converging lens. If the distance l be such as to make the two positions of the lens for the two images coincide, then $a = 0$, and $l = 4f$, the nearest distance of object and image.

201. Optical Centre of a Lens. — Let C and C' (Fig. 139) be the centres of the two spherical surfaces of the lens. Draw any two parallel radii as AC and BC'. Then the tangent planes at A and B are also parallel, and AB incident at A passes through the lens as if through a plate with plane parallel sides. There is then no deviation of the ray, its path before incidence being parallel to its path after emergence; because the interior angle of refraction equals the interior angle of incidence,

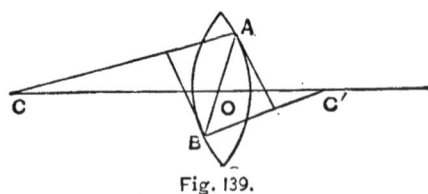

Fig. 139.

and so the external angles of incidence and refraction are also equal. Let AB be the path of such a ray through the lens. It cuts the axis of the lens in O. Then since ACO and $BC'O$ are similar triangles,

$$\frac{CO}{C'O} = \frac{CA}{C'B}.$$

Since the radii are constant in value, CO and $C'O$ are also constant, the points C and C' being fixed. O is therefore a fixed point; and all such rays as AB, whose incident and emergent portions are parallel, pass through it. O is called the *optical centre* of the lens. Its distances from the two surfaces are directly as their radii. For the proportion may be written

$$C'O : C'B :: CO : CA.$$

By subtraction

$$C'B - C'O : C'B :: CA - CO : CA,$$

or

$$\frac{C'B - C'O}{C'B} = \frac{CA - CO}{CA}.$$

But the numerators are the distances of O from the two surfaces having centres C' and C respectively, and these distances are proportional to the radii of these surfaces.

In plano-convex or plano-concave lenses the optical centre is on the convex or concave side. For if one of the radii, as $C'B$, becomes infinite so that the corresponding surface is plane, the first member of the last equation becomes zero, the denominator being infinity. The second member is also zero; but its denominator is not infinity; therefore its numerator is zero, or CA and CO are equal to each other. This can be true only when O lies on the convex surface of which C is the centre. In lenses having curvatures of both sides in the same direction, the optical centre lies without the lens.

PROBLEMS.

1, Find the geometrical focus of a small pencil of rays refracted through a double convex lens, the radii of whose surfaces are 10 and 15 cms., and the refractive index $\frac{3}{2}$, when the point-source of the light is 15 cms. from the optical centre.

2. The focal length of a lens is 20 cms., and the distance between the object and the screen 100 cms. Where must the lens be placed to give a clear image?

3. The focal length of a glass lens in air is 80 cms. If the indices of refraction of glass and water are $\frac{3}{2}$ and $\frac{4}{3}$ respectively, what is the focal length of the lens in water?

202. Construction for Images in a Converging Lens.

— (*a*) When the object is farther from the lens than the focal distance. The image is then real. Let *AB* (Fig. 140) be the object and *MN* a double convex lens, of which

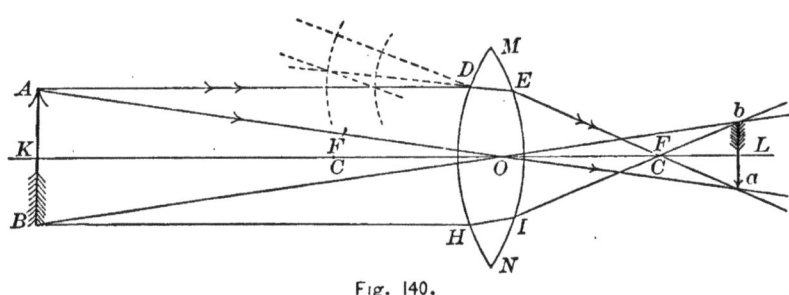

Fig. 140.

O is the optical centre. First draw secondary axes *A O* and *B O*. Since these pass through the optical centre, the rays along them undergo no deviation. Next draw *AD* and *BH* parallel to the principal axis *KO*, which passes through the centres of curvature *C*, *C*, and the optical centre. These rays may be traced after refraction by the method of (190), assuming the index of refraction from air to glass to be $\frac{3}{2}$. After refraction at the second surface both

rays pass through the principal focus *F*. It may be convenient to observe that if the index of refraction is $\frac{3}{2}$ and
r = r', the principal focus is at the centre of curvature.
Then the two rays drawn from *A* meet after refraction at
a, and those from *B* at *b*. Hence *ab* is the real image of
AB. Conversely, if *ab* were the object *AB* would be the
image. The image is inverted because the secondary axes
cross between object and image. The size of the image
is to that of the object as *LO* to *KO*.

(*b*) When the object is nearer the lens than the focal
distance. The image is then virtual, erect, and larger
than the object. Let *F* be the principal focus and *AB*
the object (Fig. 141). Proceed as before by drawing the

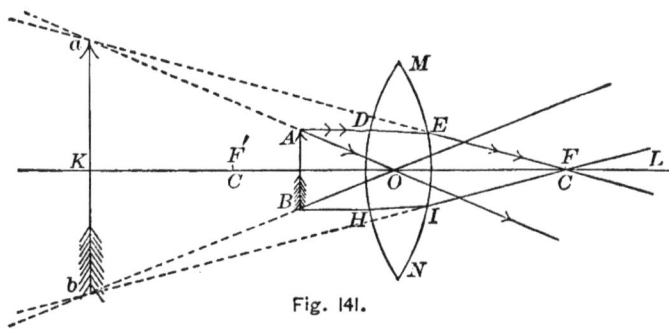

Fig. 141.

path of two rays *A O*, *B O* along secondary axes, and *AD*,
BH parallel to the principal axis, and through *F* after refraction at both surfaces of the lens. The two rays from
A diverge after emerging from the lens, but if their directions be projected backwards they will meet at *a*. So also
the two corresponding rays from *B* emerge from the lens
as if they came from *b*. Hence *a* and *b* are virtual foci
and *ab* is the virtual image of *AB*. In this case object
and image are on the same side of the optical centre.

203. Construction for the Image in a Diverging Lens.
— The images formed by concave lenses are always
virtual, erect, and smaller than the object. They may be
drawn in the same manner as those for converging lenses.

AB (Fig. 142) is the object. · The rays parallel to the

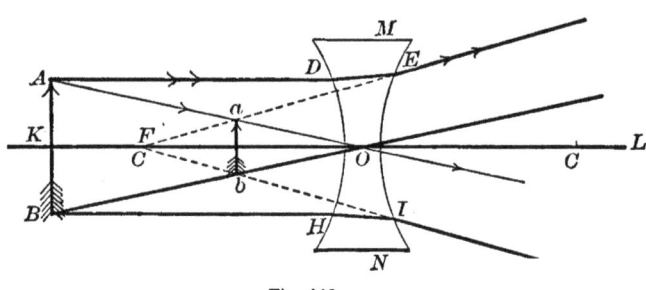

Fig. 142.

principal axis after emerging from the lens diverge from
the axis as if they came from the principal focus *F*. Their
directions produced backwards intersect the other two
rays along the secondary axis in *a* and *b* respectively.
Hence *ab* is the virtual image of *AB*.

**204. Spherical Aberration and Distortion of Image
(B., 444; P., 85).** — It must not be overlooked that the
formula which we have used to connect the object and
the image with the focal length of a lens is only an
approximate one. It was deduced by assuming the point
of incidence on the lens very near the principal axis, and
the point-source of light was supposed to be on the axis.

If the approximations made in deducing the formula
(196) are carefully examined, it will be found that the
focal length for rays incident on a lens near the boundaries
of the spherical surfaces is shorter than for rays incident
near the centre. It follows that when parallel rays are

incident on a lens the focus for the peripheral rays is nearer the lens than for the central rays. This distribution of the focus along the axis is called *spherical aberration.* Between the two foci for the outer and the inner rays the refracted beam will have a minimum cross-section, which is called the *circle of least confusion.* The rays refracted in this manner from different parts of the lens are tangent to a curve called a *caustic by refraction.*

The corrections for spherical aberration are two: First, the outer rays are cut off by an annular diaphragm. This is partly the office of the iris of the eye. Second, the surfaces of the lens are so shaped that a spherical wave before refraction remains spherical after refraction. The surfaces of the lens are then only approximately spherical.

A surface which refracts to one point the light diverging from another is called an *aplanatic surface.*

The formulæ and discussions to which this inquiry leads are tedious and intricate, and do not come within the limits prescribed for this book.

Besides the indistinctness or confusion of the image to which spherical aberration leads, there is a distortion with spherical lens surfaces. It has been assumed that a straight object will have a straight image. But with spherical lenses the image of a straight line at one side of the axis is a curved line, and the image of a plane surface is convex. From the equation

$$\frac{1}{q} - \frac{1}{p} = \frac{1}{f}$$

we obtain

$$\frac{q}{p} = \frac{f}{f + p}.$$

Now q is the distance of the image and p that of the object. Hence, while q increases with p, both measured

along secondary axes, for points of a straight object further
and further from the principal axis, yet it does not increase
in the same ratio, since $\dfrac{f}{f+p}$ diminishes as p increases.
This means that the image ab (Fig. 141), for example, is
concave toward the lens, the distance of its outer parts, a
and b, from O bearing a smaller ratio to the corresponding
distances of A and B than those of points near the axis
to the corresponding points of the object.

CHAPTER X.

DISPERSION.

205. The Complexity of White Light (V., II, 475). — When a thin beam of sunlight is passed through a prism it not only suffers deviation, as already explained, but it is resolved into a number of colors of different refrangibility. This phenomenon is known as *dispersion*.

The production of colors in this way was certainly known to Seneca, and Kepler made use of an equilateral glass prism for the study of the subject. But Newton was the first to recognize the true import of the phenomenon, and to refer the colors to the heterogeneity of white light. This was in 1666.

Admit a horizontal beam of sunlight into a darkened room through a narrow vertical slit and focus by means of a long-focus lens on a screen at a suitable distance. Then interpose in the path of the beam a glass prism (or, better, one with plane glass faces filled with carbon disulphide) with its refracting edge vertical. The beam of light will undergo deviation and dispersion, and the screen should be moved to receive it, keeping its distance from the lens the same. The order of colors is red, orange, yellow, green, blue, violet. Another color, indigo, is sometimes distinguished between blue and violet. Red is the least refrangible, and violet the most. Each color has a different refrangibility. The color is the physiological character of a light; the refrangibility is its physical character.

Such a succession of colors in the order of refrangibility, obtained from any source of light, is called a *spectrum.*

The index of refraction for red is less than for violet; and since the relative index is inversely as the speed of light in the medium, it follows that red light is transmitted through the prism with a greater speed than violet; the other colors are transmitted through transparent media with intermediate speeds. The phenomenon of dispersion is therefore due to the unequal retardation in the speed of transmission of the different colors through transparent media. Violet suffers a greater retardation or travels more slowly than red on entering an optically denser medium. Measurements of wave-length show that the ethereal undulations producing extreme violet are the shortest of all those coming within the visible spectrum. It follows that disturbances of short wave-length undergo greater diminution of speed on entering dense transparent media than those of longer period.

There is no evidence that ethereal undulations of different wave-length travel with different speeds in the ether of space.

206. Recomposition of White Light. — Newton's conclusion that refraction does not produce the colors, but serves only to separate those already mingled in white light, was confirmed by recombining the separated colors into a beam of white light. One of the methods employed for this purpose was the reflection to one point by seven mirrors of the different rays of the spectrum. The superposition of these produced white light. But a more conclusive experiment was the actual synthesis of white light from the spectral colors by means of a second prism identical with the first and placed with its refracting edge

turned in the opposite direction (Fig. 143), where S is
the incident beam which is re-
solved into the spectral colors
by the first prism and recom-
bined into an emergent beam E
of white light. •

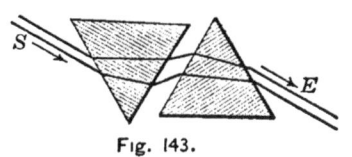

Fig. 143.

207. Dispersive Power. — If two prisms of different
materials are made with such angles that they give the
same minimum deviation for the brightest part of the spectrum, it will be found that the lengths of the two spectra
will not be the same. The angular separation of the
colors varies with the transparent medium employed. If
d', d'', d, represent the two extreme and the mean deviations in the spectrum and μ', μ'', μ, the corresponding
indices of refraction, then $d' - d''$ is the angular separation
of the two extreme colors of the spectrum, or the *dispersion*. But by Art. 194, when the refracting angle is
small, the deviation is $A (\mu - 1)$. Therefore

$$\frac{d' - d''}{d} = \frac{A (\mu' - 1) - A (\mu'' - 1)}{A (\mu - 1)} = \frac{\mu' - \mu''}{\mu - 1} = \frac{\varDelta \mu}{\mu - 1},$$

and this ratio is called the *dispersive power* of the substance. It is the ratio of the difference of deviations of
two selected rays of the spectrum to the mean deviation.
This ratio is constant for the same substance so long as
the refracting angle of the prism is small, but it is different
for different substances. Thus for crown glass the dispersive power is 0.0434, while for carbon disulphide it is
0.1466. For the same deviation, therefore, a hollow prism
filled with carbon disulphide will give a spectrum more
than three times as long as the one produced by a prism
of crown glass.

208. Chromatic Aberration (T., 113; V., II, 568). —

Since the homogeneous colors of white light have different indices of refraction, it follows from the formula

$$\frac{1}{f} = (\mu - 1)\left(\frac{1}{r} - \frac{1}{r'}\right)$$

that a single lens has different focal lengths for different colors, and that f is less as μ is greater. Hence violet light comes to a focus nearer the lens than red. Thus in

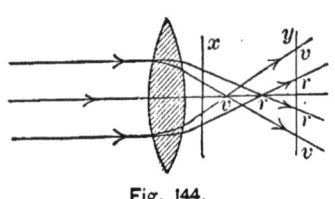

Fig. 144.

Fig. 144 v is the principal focus for violet rays and r for red. The other colors have foci lying between these two. If, therefore, a screen be placed at or near v, as at x, the image will be bordered with red; if at y, near the focus r, it will be fringed with violet. The image with least color will be obtained by placing the screen midway between the two foci where the beam of light has its smallest cross-section; but a colorless image cannot be obtained with a single lens. This confusion of colored images is called *chromatic aberration.*

Newton believed that these colors were unavoidable. According to him, the dispersion being proportional to the refraction, the one could not be suppressed without annulling the other. But Euler observed that the eye gives colorless images; and as this organ is composed of several refracting media, he concluded that one should be able to make achromatic lenses by means of two or more glass lenses united by some liquid.

Newton's experiments were shown to be inexact, and the English optician Dolland succeeded, by the use of a prism of glass and one of water with a variable angle, in making the colors disappear while retaining still a certain amount of deviation. A little later (1757) he succeeded in making an achromatic lens by a combination of crown and flint glass.

209. **Conditions of Achromatism** (T., 114; V., II, 569; D., 492; B., 452). — It will be evident from a consideration of Art. 207 that by varying the refracting angle of two thin prisms of different materials, such as crown and flint glass, and by combining them with their sharp edges turned in opposite directions, it is possible to secure *deviation without dispersion,* or *dispersion without deviation.* The conditions governing the first are those required to secure an achromatic image with two lenses; those of the second apply to a direct-vision spectroscope.

The expression for dispersion is $A \Delta \mu$. Suppose we have a second prism for which the dispersion is $A' \Delta \mu'$. Then if the image is to be colorless the dispersion of the colors in one direction must equal the dispersion in the other; that is, the angles of the prisms must be such that the dispersion shall be the same for both. Then

$$A \Delta \mu - A' \Delta \mu' = 0,$$

or

$$\frac{A}{A'} = \frac{\Delta \mu'}{\Delta \mu}.$$

Hence the condition of achromatism for the two prisms is that their refracting angles must be inversely proportional to the differences between the indices of refraction of the two pairs of selected rays near the limits of the spectrum.

But since these two prisms, which are assumed to have *different dispersive powers,* produce the *same dispersion,* the deviations produced by them cannot be the same.

For while

$$\frac{A \Delta \mu}{A (\mu - 1)} \lessgtr \frac{A' \Delta \mu'}{A' (\mu' - 1)} \quad \cdot \quad \cdot \quad (207)$$

and the numerators are equal, it follows that the deviations, represented by the denominators, are unequal. There will, therefore, be deviation without dispersion, and the deviation will be due to the material having the smaller dispersive power.

But an achromatic combination of two lenses, as well as two prisms, may be used to superpose the foci for two given colors. Suppose the system composed of two very thin lenses placed in contact. The conditions of approximate achromatism are not difficult to find. The power of a lens is the reciprocal of its focal length, and the refracting power of a system of two lenses of focal length *f* and *f'*, and very near together, is given by the equation

$$\frac{1}{F} = \frac{1}{f} + \frac{1}{f'},$$

or $$\frac{1}{F} = (\mu - 1)\left(\frac{1}{r_1} - \frac{1}{r_2}\right) + (\mu' - 1)\left(\frac{1}{r_3} - \frac{1}{r_4}\right),$$

where μ and μ' are the indices of refraction of the two kinds of glass composing the two lenses for a definite ray of one color, as, for example, red. The r's are the radii of curvature of the four surfaces.

For the other simple light, violet, for example, the indices for the two lenses being $\mu + \Delta\mu$ and $\mu' + \Delta\mu'$, the focal distance of the system becomes $F - \Delta F$. Now in order that the foci of the two selected colors may coincide ΔF must be zero. Obviously this will be the case when

$$\Delta\mu\left(\frac{1}{r_1} - \frac{1}{r_2}\right) + \Delta\mu'\left(\frac{1}{r_3} - \frac{1}{r_4}\right) = 0,$$

since these are the only terms entering into the expression for $F - \Delta F$ which differ from the expression above for F.

Since $\Delta\mu$ and $\Delta\mu'$ are of the same sign, the parenthetical quantities must be of opposite signs; that is, if one of the lenses is convergent, the other is divergent. The system as a whole will be convergent if the converging lens is composed of glass of the smaller dispersive power.

From the equation for f, viz.,

$$\frac{1}{f} = (\mu - 1)\left(\frac{1}{r_1} - \frac{1}{r_2}\right),$$

we have $\dfrac{1}{r_1} - \dfrac{1}{r_2} = \dfrac{1}{\mu - 1} \cdot \dfrac{1}{f}$; $\dfrac{1}{r_3} - \dfrac{1}{r_4}$ has a similar value.

Substituting in the equation of condition above and

$$\frac{\varDelta\mu}{\mu - 1} \cdot \frac{1}{f} + \frac{\varDelta\mu'}{\mu' - 1} \cdot \frac{1}{f'} = 0.$$

From this equation we see that in an achromatic system of two lenses their focal lengths are proportional to their dispersive powers.

The above equations may be used to determine the four radii. Generally r_3 is made equal to $-r_2$ and the other two are arranged to make the spherical aberration as small as possible. Fig. 145 shows a combination of flint and crown glass for achromatism. Since the dispersive power of crown glass is only about half that of flint (0.0434 and 0.0753) the converging lens is of crown glass and the diverging of flint for a converging combination. The outer surface of the flint glass is not usually plane, nor are the radii of curvature of the two surfaces of the crown glass equal to each other for the best definition.

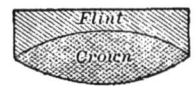

Fig. 145.

PROBLEMS.

REFRACTIVE INDICES FOR THE FRAUNHOFER LINES.

	A	B	C	D	E	F	G	H
Crown Glass . .	1.5089	1.5109	1.5119	1.5146	1.5180	1.5210	1.5266	1.5314
Flint Glass . . .	1.6391	1.6429	1.6449	1.6504	1.6576	1.6642	1.6770	1.6886
Water	1.3284	1.3300	1.3307	1.3324	1.3347	1.3366	1.3402	1.3431
Carbon Disulphide,	1.6142	1.6207	1.6240	1.6333	1.6465	1.6584	1.6836	1.7090

1. Find the dispersive power of crown glass and flint glass for the lines *A*, *H*, and *D*.

2. If the refracting angle of a crown-glass prism is 20°, what must be that of a flint-glass prism to produce achromatism, and what will be the resulting deviation for *A*, *H*, and *E*?

3. Suppose we wish to reunite the colors represented by the *B* and *G* lines in crown and flint glass; what must be the relative focal lengths of the two lenses of the combination for the *E* line?

210. Irrationality of Dispersion (B., 452; P., 191; D., 493). — If the dispersion or angular separation of two colors be made the same for prisms of any two media, the dispersion for the other colors of the spectrum will not in general be the same. While the order of the colors in the two spectra is the same, except for a few abnormal cases, yet the relative spaces occupied by the several colors are not the same. This is known as the *irrationality of dispersion*. Thus the spectra produced by prisms of crown and flint glass may be made of the same length by properly adjusting the refracting angles, but the lengths occupied by the same colors in the two will not be the same. For crown glass orange and yellow are spread over a relatively greater area than for flint glass. The relative distribution of the colors throughout the refraction spectra of different transparent media is not the same.

Obviously, therefore, the achromatism secured by combining two lenses must be of an imperfect kind. Correction for the extreme colors does not correct for the intermediate ones; and there remains, therefore, a residuum of color. Only two colors can be superposed by a double combination of lenses. To superpose three colors a triple combination is necessary. For a double combination rays of extreme refrangibility are not generally chosen, but the orange-yellow and the greenish-blue.

211. Dark Lines in the Solar Spectrum. — A spectrum consists of a succession of colored images of the slit. If this slit is not extremely narrow the several images will overlap, giving what is called an impure spectrum. To obtain a pure spectrum the slit must be narrow and an achromatic lens must be placed at a distance from the slit equal to its focal length, so that the light emerging from it may consist of parallel rays. These parallel rays may be received upon a prism adjusted for minimum deviation. The resulting pure spectrum should then be viewed with an achromatic telescope. A piece of apparatus for obtaining and viewing a pure spectrum is called a *spectroscope.* When it is provided with a divided circle for measuring deviations it is called a *spectrometer.* A simple form of spectrometer is shown in Fig. 146. At the left is an adjustable slit placed at the principal focus of the lens contained in the other end of the same tube. Parallel rays pass from this tube to the prism and are there refracted so as to pass into the observing telescope on the right. This

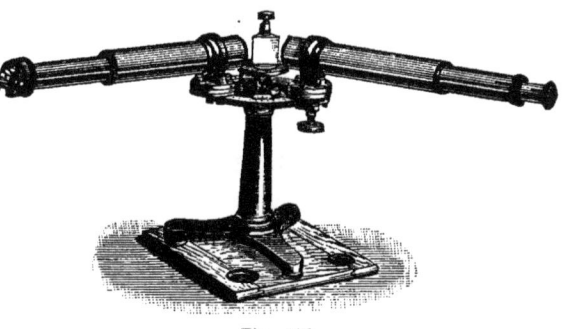

Fig. 146.

is movable so as to take successive parts of the spectrum into the field of view.

When the solar spectrum is examined by such a spectroscope it is found to be crossed by numerous dark lines. These dark lines were first observed by Wollaston in 1802 by looking at a slit in the shutter of a dark room through a prism held in the hand at a distance, and with its refracting edge parallel to the slit. The rays reaching the prism were then nearly parallel. Fraunhofer studied them in 1814–15, counted about 600 of them, and marked the places of 354 on a map of the spectrum. They are therefore often called Fraunhofer's lines. The solar spectrum is discontinuous, or is characterized by the absence of numerous colored images of the slit. These dark lines represent the places of rays of definite refrangibility or wave-length. Some of them are always present in the solar spectrum. These have their origin in the sun itself. Others are greatly strengthened or appear only as the sun nears the horizon. These variable lines have their origin in the earth's atmosphere, and are called atmospheric or telluric lines.

Fraunhofer designated the chief lines mapped by him *A*, *B*, *C*, *D*, *E*, *F*, *G*, *H*. He afterwards added the line *a* in the red and the line *b* in the green. These dark lines extend from *A* in the extreme red to *H* in the extreme violet.

212. Three Kinds of Spectra. — When the spectra from different sources of light are classified they fall into three groups, each of which may have several subdivisions:

(*a*) *Bright-line spectra.* The spectra of incandescent gases and vapors consist of a limited number of *bright lines*, each of which is a monochromatic image of the slit — an image in one color. Thus the spectrum of sodium vapor is the yellow line marked *D* on the maps. With a spectroscope of sufficient resolving power this yellow line is found to consist of two lines, and each one of these is double. Other vapors produce other colored lines, and no two vapors or gases give rise to the same series of bright-line images of the slit.

(*b*) *Continuous spectra.* Next in the order of complexity come the continuous spectra of incandescent or white-hot solids and liquids. They exhibit a perfectly continuous succession of colors from one extremity of the spectrum to the other without any interruptions or gaps. The spectra of lights used for artificial illumination, such as those of candles, gas flames, or the electric light, are continuous. Their light is due chiefly to white-hot carbon, an incandescent solid. The extension of these spectra toward the more refrangible or violet end depends upon the temperature of the source.

(*c*) *Absorption spectra.* These are discontinuous, like the solar spectrum, and they are rendered such by losses due to absorption in the passage of light through transparent media. The absorption producing the dark lines of the solar spectrum takes place chiefly in the outer envelope of the solar atmosphere. The incandescent mass of the sun, which would by itself present a continuous spectrum, is surrounded by an atmosphere of gases and vapors at a high temperature, traversing which are the rays emitted by the photosphere. Absorption takes place in this reversing layer; and the dark rays produced, which are only relatively dark in comparison with the adjacent bright portions of the spectrum, are the inversion of those luminous rays which form the emission spectra of these gases and vapors in the outer envelope of the sun.

The principle of absorption is the same as that of resonance or co-vibration in sound. Every gas or vapor when white hot emits rays of the same wave-length as those which it absorbs from an independent source when at a lower temperature. Thus the *D* line of the solar spectrum coincides exactly with the bright line given by sodium vapor in a state of incandescence. Not only has the *coincidence* been established between the Fraunhofer *D* line and the yellow line of sodium vapor, but the *reversal* of this yellow line by sodium vapor as the absorbing agent has been accomplished. These results laid the foundation for the science of spectrum analysis by which the approximate chemical composition of self-luminous celestial bodies has been made out.

Kirchhoff in 1860 established the following law of spectrum analysis :

The relation between the emissive power and the absorbing power, relative to any definite radiation, is the same for all bodies at the same temperature. If light from a vapor at a higher temperature traverses the same vapor at a lower temperature, the light absorbed in the latter is greater than the light emitted by it. The result is relatively dark lines or a reversal.

CHAPTER XI.

INTERFERENCE AND DIFFRACTION.

213. Interference of Light from two Similar Sources (T., 185; B., 488; P., 119). — The phenomena resulting from the superposition of two systems of waves of homogeneous light, travelling in nearly the same direction, are called *interference*. Similar phenomena have already been discussed in sound.

When the ether at any point is affected simultaneously by two waves it is thrown into vibration by both, and the result is a compound motion into which each vibration enters independently. If the two vibrations are in the same direction, their amplitudes are added and the resulting amplitude is the sum of the two components; but if they are of opposite sign the resultant amplitude is equal to the difference of the constituent amplitudes. If in this latter case the amplitudes of the two vibrations are equal the two motions should completely annul each other.

We know already that two systems of sound-waves may interfere so as to produce silence, and two water-waves may be superposed so as to leave the surface undisturbed. If the undulatory theory of light is true it should be possible to add light to light in such a manner as to produce darkness. This actually occurs in many cases. The limits set to this book will enable us to explain only a few of the most simple ones.

The most simple arrangement to exhibit interference

was devised by Fresnel. BCD (Fig. 147) is an isosceles glass prism with the angle at C nearly 180°. Let O be a source of homogeneous light in a plane through the angle C of the prism perpendicular to BD. The light passing

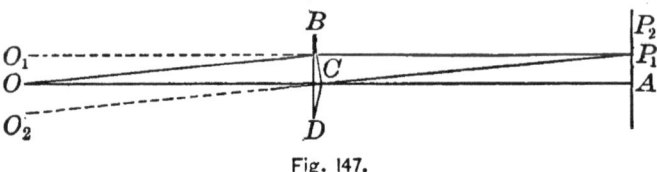

Fig. 147.

through the prism will consist of two parts diverging from the virtual sources O_1 and O_2. Since the point A is equidistant from these sources, the two systems will arrive at A in the same phase and will reënforce each other. Hence on the screen at A there will appear a bright band parallel to the refracting edge of the prism.

Let P_1 be so situated that the distance $P_1 O_2$ shall exceed $P_1 O_1$ by half a wave-length for the homogeneous light employed, the yellow of sodium, for example. Then the two systems will meet at P_1 in exact opposition of phase. This is the condition for destructive interference, and at P_1 there will be a dark band parallel to the bright one through A.

At P_2 where $P_2 O_2$ exceeds $P_2 O_1$ by an entire wavelength there will be a bright band again; and still further out will be found a second dark band, and so on. We therefore conclude that if $O_2 P - O_1 P = n\dfrac{\lambda}{2}$, where λ is the wave-length of the light employed, there will be a dark band for all odd values of n, and a bright band for even values. The screen will then be illuminated with a series of bright bands alternating with dark ones. By measuring the distance from A to the first dark band,

and the distance AO it is possible to compute the differ-ence $P_1O_2 - P_1O_1$, and so to measure the wave-length of the light. A measurement of this kind shows that the wave-length of yellow sodium light is only 5.89×10^{-5} cm. Hence the vibration-frequency, which is the quotient of the speed by the wave-length, is about 500 millions of millions a second.

Particular attention should be given to the fact that there is no loss of energy in interference, but only a redis-tribution of it among the dark and light bands.

214. Interference produced by Thin Films (T., 195 ; D., 503 ; P., 138 ; B., 492). — Interference phenomena are produced by thin transparent films. The iridescence of ancient glass, of a thin film of oil on water, of oxide on the surface of polished or molten metal, and of the soap bubble are familiar examples. These colors are the residuum of white light left after some portions have been cut out by interference between the two wave systems reflected from the parallel surfaces of the film.

Let AA (Fig. 148) be one sur-face of the film and BB the other. Light incident on the upper surface is partly trans-

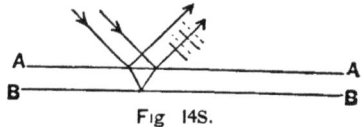

Fig 148.

mitted and partly reflected. The portion penetrating the first surface is in part reflected from the second, and emerges again from the first surface along with light which has undergone reflection only. If now the thickness of the film be such that the internally reflected system falls behind the other by a whole wave-length of the light em-ployed, then this system will on emergence be in the same phase as the system reflected externally, so far as difference of path is concerned ; and if difference of phase depends only on difference of path traversed, when this difference vanishes with a film of infinitesimal thickness · the two pencils should be in the same phase and the illumination a maximum. But the fact is when a film is made as thin as possible it becomes black. All the light reflected is extinguished by interference. The difference of phase therefore depends on something besides difference of path. This is found in the fact that the two reflections take place under different conditions, one in the rare medium next to the dense, and the other in the dense medium next to the rare. One of the two interfering systems in consequence loses half an undulation relative to the other in the mere act of reflection. This is analogous to the change in phase of a sound-wave reflected from the end of an open organ pipe. When a condensation is reflected from the closed end of a pipe there is a change of sign of the motion, but not of the condensation. When it is reflected from the end of an open pipe, there is no change of sign of the motion, but the condensation changes sign, or is reflected as a rarefaction. Therefore if two condensations could be brought together after reflection, one from the closed end of a pipe and the other from an open end, the two disturbances would be found in opposite phases and would interfere. So two pencils of light reflected under the corresponding opposite conditions have thereby impressed upon them a difference of phase equal to half a period.

When therefore the thickness of the film and the angle of incidence are such that one pencil falls behind the other in transmission by a whole number of wave-lengths, interference takes place with extinction of light, since a phase difference of half a period must be added because of the reflection at the two surfaces under opposite conditions.

With white light the extinction of one spectral color by

interference leaves colored fringes. It is to be observed that the thickness of film that would produce a relative retardation of the internally reflected pencil of one wave-length for violet would be a retardation of only about half a wave-length for red (217). Therefore extinction of both colors cannot take place at the same part of the film. If the violet is cut out the red remains. Similar reasoning applies to the intermediate colors. The reflected light is therefore fringed with color.

215. Diffraction Fringes with a Narrow Aperture (A. and B., 443; P., 175; B., 501). — When a beam of sunlight is admitted through a very narrow slit into a darkened room, and is received upon a screen at some distance, there will be seen a central band of white light in the direct path of the beam, bordered with colored fringes. Through so small an opening light passes not merely as a definite pencil, but it also diverges in all directions from all points of the opening as new centres of disturb-ance. This phenomenon is called *diffraction*. The waves of light thus bend around an obstruction, like the edge of the slit, in the same manner as water-waves run around an obstruction. The colors will be intensified if, instead of a single narrow opening, a series of equidistant fine lines ruled on smoked glass be employed. This *grating*, as it is called, should be placed at the focus of a lens through which a beam of sunlight passes. The room must be well darkened, and

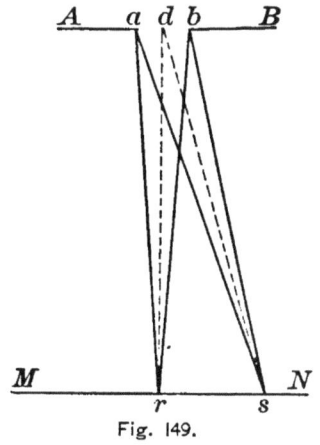

Fig. 149.

all extraneous light should be cut off by properly arranged screens.

Let *ab* (Fig. 149) be a section of the slit, the length of
which is perpendicular to the plane of the paper, and let
MN be the screen. Then *r* in the direction of the pencil
is nearly equidistant from every point between *a* and *b*,
and all the secondary waves, starting from the several
points between *a* and *b* as centres, arrive at *r* in nearly
the same phase. There is, therefore, maximum illumination at this point. Let *s* be a point so situated that the
distance *as* shall exceed *bs* by one wave-length λ for
some color, as red. Then red will suffer extinction by interference at *s*. For if *ab* be divided into two equal parts
the difference in distance of *a* and *d* from *s* is half an
undulation, or one-half λ; the secondary waves from *a*
and *d* meet at *s* in opposite phases; and every wave
from a point between *a* and *d* meets at *s* a wave of opposite
phase from a corresponding point between *d* and *b*.
Therefore total extinction of light of this particular wave-
length λ takes place at *s*.

If the distances of *a* and *b* from a point differ by three
halves λ, then so far as the illumination at this point is
concerned the slit may be divided into three equal parts;
the secondary waves from two of these parts interfere
at the screen as explained, while those from the third part
produce illumination.

In general if *as* — *bs* on either side of *r* is an even
number of half-waves there will be an even number of
half-period elements in *ab*, which will mutually interfere
at *s*, and the illumination will be less than if there is an
odd number of half-period elements in *ab*. With monochromatic light, bright and dark bands will alternate on
either side of *r*. The position of these bands is given

by the equation

$$as - bs = n\frac{\lambda}{2},$$

where n is even for the dark and odd for the bright bands.

With white light, in which λ is different for the different colors, extinction for different colors will take place at different distances of s from r, and hence colored fringes will appear on the screen.

216. Spectrum by a Diffraction Grating (A. and B., 446; P., 186). — A system of very narrow, equal, and equidistant rectangular apertures is called a *diffraction grating*. These may be made by cutting with a diamond point by means of a dividing engine a number of parallel equidistant lines on a glass plate. The light then passes through the transparent spaces between the lines.

Let a plane wave approach the grating in the direction of the arrow (Fig. 150). Let a, c, etc., be the parallel apertures, and let parallel lines ab, cd, etc., be so drawn that the distance ae to the foot of the perpendicular from c on ab shall equal some definite wave-length of light λ. Then an will be an exact number of wave-lengths $n\lambda$, co will be $(n-1)\lambda$, and so on. The line mn will therefore touch the front

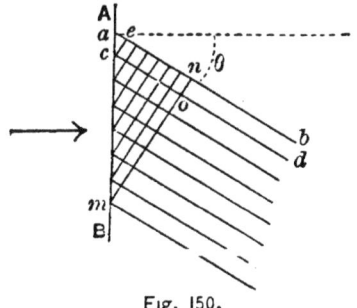

Fig. 150.

of a series of secondary waves from the successive transparent openings, all in the same phase; and if this new plane wave be transmitted through a lens with its axis parallel to ab, all the transparent apertures of the grating will send light to the focus in the same phase and of wavelength λ.

Let θ be the angle between ab and the direction of the light incident on AB. Then

$$ae = \lambda = d \sin \theta,$$

where d is the distance ac from centre to centre of the openings. The disturbance at the focus of the lens will be the resultant of all the disturbances coming from the numerous apertures of the grating, and all those of wavelength $d \sin \theta$ will arrive in the same phase.

For disturbances of any other wave-length this coincidence in phase will not exist. If, for example, the difference between λ and $d \sin \theta$ is $\frac{1}{100}\lambda$, then the light from the first opening will be opposite in phase to that from the fifty-first, that from the second will be in opposition to that from the fifty-second, etc. Hence all the light from the first fifty openings will exactly neutralize that from the second fifty in the direction θ. Since the number of lines on the grating is several thousand to a centimetre, light of only one wave-length is found at any angle θ with the direction of the incident beam. Hence a pure, *normal spectrum* is produced in which with a given grating the angular separation of the different colors depends only on their wave-length.

From the formula $\lambda = d \sin \theta$ it is obvious that the longest waves are found at the greatest deviations; for the first spectrum the wave-lengths are nearly proportional to the deviations, for θ_1 is small. Hence there can be no irrationality of dispersion in a diffraction spectrum.

For greater deviations we may have

$$2\lambda = d \sin \theta_2.$$

Such waves produce a spectrum of the second order lying farther away from the normal to the grating. For the spectrum of the third order

$$3\lambda = d \sin \theta_3,$$

and for the nth spectrum

$$n\lambda = d \sin \theta_n.$$

These spectra are all violet at the inner edge nearest the normal to the grating and red at the outer. Overlapping of spectra will occur when the deviation for the violet of any order is less than that of the preceding red. Moreover, since the deviation is approximately proportional to wave-length, and the wave-length of red is somewhat less than twice that of violet, the red of the second spectrum will overlap the violet of the third. The separation of the superposed parts may be effected by means of a prism.

The diffraction grating·furnishes an admirable method of measuring wave-lengths. The distance d, which is a constant of the grating, must be measured, and the spectrometer must be supplied with a graduated circle for the purpose of measuring the deflection θ, corresponding to any dark line of the solar spectrum, or to any bright line of an artificial spectrum of a vapor.

Spectra similar to the preceding may be obtained by reflection from a grating ruled with very fine parallel grooves on speculum metal. The surface must first be finely polished. The pencils reflected from the polished intervals between the rulings come as if from virtual images of the source through the intervals of the grating. The exquisite colors of mother-of-pearl and other striated surfaces, of the feathers of certain birds, and of changeable silk are instances of the same method of producing colors from white light by diffraction. ·

Professor Rowland's famous reflecting gratings are ruled on concave surfaces of polished speculum metal.

217. Wave-lengths and Vibration-frequencies. — The unit employed in measuring wave-lengths of light is the *tenth-metre*, of which 10^{10} are required to make a metre. The following are the values for the principal Fraunhofer lines in air at 20° C. and 760 mm. pressure:

A	7621.31	E_1	5270.52
B	6884.11	E_2	5269.84
C	6563.07	F	4861.51
D_1	5896.18	G	4293.
D_2	5890.22	H_1	3968.

Taking the speed of light as 300 million metres a second, or 300×10^{16} tenth-metres, the vibration-frequencies corresponding to the above spectral lines may be found by dividing this speed by the several wave-lengths, since

$$n = \frac{V}{\lambda}.$$

The result is as follows:

A	393.6×10^{12}	E_1	569.2×10^{12}
B	435.8 "	E_2	569.3 "
C	457.1 "	F	617.1 "
D_1	508.8 "	G	698.8 ··
D_2	509.3 "	H_1	756.0 ··

Thus the light entering the eye and producing the violet color represented by H_1 is due to 756 millions of millions of vibrations a second. A photograph of the sun has been taken with an exposure of only one twenty-thousandth of a second. But in this time a beam of light 15,000 metres (9.32 miles) in length has entered the camera, and fully 375×10^8 or 37,500 millions of waves have impressed their effects on the sensitized plate.

The visible spectrum lies between wave-lengths of about 7500 and 3900 tenth-metres. Rowland has measured them from 7714.657 to 3094.736 tenth-metres, and has photographed them from about 7000 to 3000. Langley has measured lunar radiations with wave-lengths of 170,000 tenth-metres, or nearly twenty-three times as long as the longest waves exciting vision. The invisible spectrum extends several times the length of the visible spectrum beyond the extreme violet; so that the entire invisible spectrum actually explored is perhaps thirty times as long as the visible one. Physically the only difference existing among these radiations is one of wave-length. All of them represent energy which is converted into heat when absorbed by the proper surfaces; and perhaps all may be able to excite or precipitate chemical changes if the proper sensitive substances are found for different parts of the spectrum. The mechanism of the eye limits its receptivity to the visible spectrum. The differences formerly supposed to exist between the so-called light, heat, and actinic rays are therefore differences in the receptive apparatus.

CHAPTER XII.

COLOR.

218. Modes of producing Color. — Color has no objective existence, but is the physiological character assigned to light by sensation. The only physical differences corresponding to different colors are differences in wavelength. Color in light corresponds to pitch in sound, with this difference, that an indefinite number of color mixtures may produce the same effect on the eye, while the ear analyzes a complex tone into its elements and recognizes the intervals. Hence each combination of sounds produces its own effect.

Red is due to the longest ether-waves exciting vision, and violet to the shortest. The production of colors from white light involves, therefore, some process of isolating vibrations of certain definite periods. These processes are three in number:

(*a*)   Refraction.
(*b*)   Interference.
(*c*)   Absorption.

The analysis of white light into its component colors by means of the first two methods has already been described. It remains to explain briefly the third.

219. Color of Opaque Bodies (L., 180; T., 147; P., 384). — All bodies, except those with highly polished surfaces, reflect light by irregular reflection from greater or less depths within them. If all the components of white light are reflected in the same proportion the body appears white or gray. Such is the case with a sheet of white paper or a white screen on which the solar spectrum is projected. These surfaces reflect diffused light in all directions and without preference for light of particular wave-lengths. Hence all the colors of the spectrum on such a screen appear the same as when they are received directly into the eye placed in the path of the diverging beam from a prism. But if the body exhibits any inequality in its relative absorbing and reflecting power for light of different wave-lengths lying within the visible spectrum, then it will appear colored when white light is incident upon it, the color being the result of mixing those color components of white light which the body reflects. The other spectral colors are absorbed. A mixture of these would produce a color *complementary* to that of the reflected components. Complementary colors are those which, added together, produce the impression of white light.

The colors of natural objects are therefore chiefly residuals left after absorption.

It is easy to justify this conclusion by appropriate experiments. Let a solar spectrum, with rather wide slit, be projected on a white screen by means of a carbon disulphide prism. Take a flower which shows rich red petals in large masses, such as a tulip, or certain geraniums. Hold it in the spectrum and pass it through the different colors. In the red the flower shines with its usual bright red color; but as it is moved along into the green it becomes black and continues to show no power of reflection for the remainder of the spectrum. All the other colors except red are readily absorbed. The red, on the contrary, is reflected and gives the color to the body. We see therefore that the body can exhibit no color not already present in the light which illuminates it. A piece of red flannel is brilliantly red in the less refrangible end of the spectrum, but suddenly turns to a dirty brown, and then a dead black, when moved out of the red toward the violet. Ribbons of various colors carried along through the spectrum give very instructive results, which are readily explained by the power of selective absorption possessed by them.

The essential nature of the colors of objects may be indicated by saying that they are the residue of the light, by which they are illuminated, after abstraction of the rays extinguished by absorption.

With homogeneous illumination differences of color are no longer possible. This fact is strikingly exhibited by viewing objects of various colors in a room lighted only with burning sodium. The most healthful face presents an ashen hue, and brilliant flowers are reduced to a faded yellow. " Were the sun a sphere of glowing vapor of sodium, all terrestrial nature would present this monotonous and gloomy aspect. It requires the white light of the sun, in which innumerable colors are blended, to disclose to our eyes the variegated tints of natural objects."

220. Color of Transparent Bodies (T., 151; P., 379; L., 172). — If a transparent body absorbs all radiations within the visible spectrum in equal proportion, it is colorless ; but if it is transparent to certain radiations affecting the eye and not to others, it appears colored by transmitted light, and the color is due to the mixed impression produced by the transmitted radiations.

The different colors of transparent bodies result from their individual capabilities of selective absorption. This generalization is easily justified by means of the spectrum.

Plants owe their color to the chlorophyll contained in their cells. An alkaline solution of this coloring matter, placed in the prismatic beam, produces a deep black band in the middle of the red, only feeble absorption striæ in the yellow and green, while the entire indigo-violet part of the spectrum from about the middle onward is entirely absent. The light transmitted by chlorophyll is therefore red and predominantly green.

If a piece of blue cobalt glass be interposed in the solar beam, the spectrum will consist of a small amount of the extreme red and all of the indigo-violet part which the chlorophyll absorbs.

A piece of glass, colored red with the sub-oxide of copper, allows only the red and orange-red rays as far as the *D* line to pass through. All the rest of the spectrum is completely stopped by this glass. If now the light be passed through the cobalt-blue and the copper-red glasses in succession, the only part of the spectrum surviving the double absorption process is the extreme dark red transmitted by both.

A solution of potassium bichromate transmits the less refrangible part of the spectrum only. Its spectrum stops at the Fraunhofer line *b*. A solution of the ammoniated oxide of copper transmits only the more refrangible part of the spectrum from the *b* line on. These two colors, therefore, contain all the spectral tints, and are complementary to each other. But if the two solutions in flat glass cells be placed in the path of a beam of sunlight, one behind the other, the combination scarcely permits the passage of any light. The one fluid looked at through the other appears almost perfectly black. The light that struggles through the one is stopped by the other. It must not be supposed that this experiment illustrates a mixture of colors or colored lights. Far from it! It illustrates successive absorption by transparent bodies.

If the blue ammoniated copper oxide solution be placed in front of the yellow solution of normal potassium chromate of the proper density, the light transmitted by the two will be green. If their separate spectra be examined, it will be found that both solutions transmit green. Hence green is the only color common to the two, and is therefore the only one which is not stopped by absorption in the one solution or the other.

The same explanation applies to the *green* color obtained by mixing yellow and blue pigments. The light penetrates below the surface of thin layers of such pigments and suffers absorption during transit through them. Hence green is the only color which survives the double process of absorption. A mixture of pigments is not a mixture of colored lights. It is rather a process of successive absorption; the resulting color is the residue.

221. Mixing Colored Lights. — In order to perceive the mixed effect due to two or more colors, it is necessary that they fall upon the retina either simultaneously or in quick succession. Visual impressions persist for a small fraction of a second; and, if one remains till the arrival of another, both impressions are simultaneously present.

If two partially overlapping discs of light be projected on the screen and transparent colored bodies be placed in the path of the two beams, the light reflected to the eye from the overlapping area will consist of a real mixture of the two colored beams. Thus if the ammoniated copper oxide solution be placed in the path of one beam and the potassium chromate in the other, the area common to the two discs will be white or gray with the proper density of the two solutions. When their colored images are *added* they cannot in any way be made to produce green.

So the cobalt blue and the oxide of copper red glasses will give beams of light which by addition produce white.

If a disc of cardboard, colored in sections, be rapidly rotated, the result is a mixture or superposition of visual impressions. But very different mixtures may produce the same visual impression. The eye has no power of analyzing light into its constituents. "Any one of the elementary colors, from the extreme red to a certain point in the yellowish green, can be combined with another elementary color on the other side of the green in such proportion as to yield a perfect imitation of ordinary white." The unaided eye can tell nothing about the composition of colored light; it must be studied by means of the spectroscope, armed with a prism or a diffraction grating.

222. Three Primary Color Sensations (B., 479; D., 529). — The Young-Helmholtz theory of color sensations supposes that each element of the retina, broad enough to perceive light, consists of three ultimate nerve-ends, each of which serves to give perception of one of three physiologically primary colors. All other color perceptions are due to the simultaneous excitation of these three sets of nerve-ends in varying relative degrees. The three primary color sensations are red, green, and violet, though apparently any three such colors may be made the basis of a systematic classification of colors.

The theory supposes that one set of nerve terminals gives **red** when excited by waves of long period; the second gives green when excited by waves of intermediate period; and the third gives violet when excited by waves of short period. When the first and second set are both stimulated the resulting sensation is yellow; when the second and third are alone stimulated the result is a sensation of blue. An equal stimulation of all three produces the impression of white light.

Hence red and green light mixed produces on the eye the impression of yellow, while the spectroscope shows at the same time the entire absence of the spectral color yellow. Yellow as a sensation may therefore be produced either by one stimulus of a definite period, which excites both the first and second sets of nerve-ends; or by two stimuli, corresponding to red and green, exciting the two sets of nerve-ends simultaneously. So green and violet give blue. In general colors near each other in the spectrum give when compounded an intermediate color sensation.

Since yellow contains both the red and green sensations or stimuli, and blue contains both the green and violet, when yellow and blue are compounded they contain all three color sensations, and produce the impression of white.

The visual apparatus of the eye is not affected by radiations below the extreme red, nor by those above the extreme violet. The blindness of the eye to very long and very short ether-waves may be an advantage. The energy of the former is so great that ordinary vision would be impossible if our eyes responded to their stimulus; while if the ultra violet rays produced vision chromatic aberration would be excessive, and clear images would be impossible. What spectral colors would be added to the chromatic scale by an extension of vision to either the infra red or the ultra violet waves we have no means of conjecturing.

223. Subjective Colors. — The theory of primary color sensations furnishes a ready explanation of colors due to fatigue of the retina. It is well known that objects are quite invisible to one entering a faintly lighted room after exposure of the eyes for a little time to a bright light. After being subjected to a ₛtrong stimulus the eye loses its sensitiveness to a weak one. This liability to fatigue is characteristic not only of the retina as a whole, but of any portion of it giving one of the primary color sensations. Fatigue of the retina causes it to lose the power of responding to the stimulus of any color long looked at ; and when the eye is then directed toward a moderately illuminated white surface it appears of a tint complementary to the one which has produced the fatigue. This is one form of subjective colors. If the eye be fixed for half a minute on a colored picture, red, for example, in a strong light, and be then directed to a less strongly illuminated white wall, an image of the picture will be seen in green, enlarged if the wall is more distant than the picture itself. The eye, fatigued for red, still maintains its sensitiveness for green. Hence the relatively faint white light is sufficient to stimulate for green while it furnishes insufficient stimulus for red. The state of the retina therefore serves to abstract from white light a sensation due to a portion of its component radiation, leaving the rest. This is a fourth method of color production, but, unlike the other three, it is not objective. It belongs more especially to physiology than to physics.

It has been shown that when the eye is fatigued with white light it recovers its sensibility for different colors successively after different intervals of time. If one looks out of a small window, such as a porthole, on a strongly illuminated sky, and then closes one's eyes, the after image of the window will appear in dissolving colors of brilliant hues.

Simultaneous color contrasts are another form of subjective colors. These are well displayed by laying thin tissue paper over black letters printed on a ground colored green. The letters in a strong light are pink by contrast. The tissue paper furnishes a faint illumination in contrast with the strong green, and the unfatigued nerve terminals giving red cause the letters to appear red or pink in contrast with the complementary green. Hence certain colors are heightened by contrast, particularly complementary ones.

CHAPTER XIII.

POLARIZED LIGHT.

224. Polarization (P., 232; D., 476; S., 1; L., 293). — In the study of light up to this point there has been no occasion to inquire respecting the direction of vibration in a wave of light. In sound the particles of the atmosphere have a longitudinal motion, and many of the phenomena of wave-motion in sound are helpful in the study of analogous phenomena in the theory of light, irrespective of any differences growing out of longitudinal as opposed to transverse vibrations.

But we now approach a class of optical phenomena of the highest interest, to which the theory of sound furnishes no parallels. Such phenomena belong to what is known as *polarized light.*

If a plate of red, brown, or green tourmaline, cut parallel to its optic axis, is held so that a beam of light falls upon it normally, the transmitted beam will be found to have undergone a most remarkable change. To the unaided eye the transmitted light differs in no respect from the incident beam, except that it has undergone a slight change in color on account of the natural tint of the particular tourmaline crystal used.

But if the transmitted beam be examined by means of a piece of plate glass, it will be found that while in one direction it is reflected in the same manner as common light, yet when the glass is turned around the beam as an axis the light is found to vary in intensity; and in one position the reflected light vanishes entirely. The light transmitted by the tourmaline can be reflected in one plane at all angles, but in the plane at right angles to this it is imperfectly reflected, and at a certain angle of incidence is not reflected at all.

Further, if the beam transmitted through one crystal of tourmaline be examined by a second parallel plate, cut in the same way, it will be found that in one position of the second plate the light is freely transmitted, while it becomes feebler and feebler if either plate be turned around in its own plane; and when the two parallel plates have their longer dimensions at right angles, no light whatever passes through them. The light is completely extinguished by crossing two transparent and nearly colorless crystals. The beam of light transmitted by the first plate possesses a kind of *two-sidedness*, analogous to the *two-endedness* of a magnet; hence the analogous, though unfortunate, name of *polarization*, which is applied to it.

Such a beam of light is said to be *plane polarized*. The plate of tourmaline is called the *polarizer;* and the plate glass reflector, or the second tourmaline, the *analyzer*. The one brings the light into the condition of polarization and the other serves to examine it. There are many processes, some natural and others artificial, by means of which light may be polarized; and the apparatus serving as a polarizer may always be used as an analyzer.

It is important to observe that the proportion of the beam transmitted by the polarizer, which also passes through the analyzer, depends upon the orientation of the second plate with respect to the first, and when the two plates are crossed the beam transmitted by the first is completely extinguished by the second. Now, if the vibrations were longitudinal, it is impossible to conceive how the crossing of the plates could stop the light, since the rotation of the second plate could not affect them. The phenomenon of extinction is therefore held to demonstrate that the vibrations of the ether constituting light (and radiant energy in general) are transverse.

In plane polarized light the vibrations are confined to a single plane. Light may also be *circularly polarized*. The motions of the ether particles are then confined to circles about the ray as an axis, successive particles having equal difference of phase.

If the ether particles rotate in ellipses about the ray as an axis, the light is said to be *elliptically polarized*.

The relation of circular and elliptical polarization to two systems of plane vibrations, at right angles to each other, will be readily understood when the composition of simple harmonic motions at right angles and of the same period is recalled (Art. 130).

225. Double Refraction in Iceland Spar (D., 509; B., 510; P., 236; S., 17; L., 282; T., 205). — In the study of refraction up to this point it has been assumed that light travels with one velocity in the denser transparent medium, and that there is consequently but one refracted beam for each incident one. But all crystalline substances, except those whose fundamental form is the cube, possess the property of producing two refracted beams. This property is called *double refraction*. It belongs also to animal and vegetable substances having a regular arrangement of parts, and to transparent media, like glass, unequally strained in different directions.

If a crystal of Iceland spar (crystallized carbonate of

calcium) be laid on a printed page, the letters will in general appear double ; or if a beam of sunlight be admitted through a small round opening in the shutter and focused on a screen, and if a crystal of Iceland spar be placed in the path of the beam so that the beam is incident normally on its surface, two equally illuminated discs will appear on the screen. The incident beam is divided into two by the process of double refraction.

One of these beams as it emerges from the spar follows precisely the same course as if it had traversed a piece of glass, and is therefore called the *ordinary ray.*

The other beam, on the contrary, is laterally displaced in a direction depending on the position of the crystal; and if the crystal be rotated, keeping the angle of incidence constant, the spot belonging to the ordinary ray will remain fixed, while the other one will describe a circle around it. Further, if the angle of incidence be changed, the refraction of the second beam does not follow the law of sines. Hence this one is called the *extraordinary ray.*

In every doubly refracting crystal there is at least one direction in which no bifurcation of the ray takes place. This direction is called an *optic axis* of the crystal. The refracted rays diverge most widely when the incident beam is perpendicular to the optic axis. In Iceland spar the ordinary ray, which follows the law of single refraction, has an index of refraction for the D line of 1.6585. The extraordinary ray does not in general lie in the plane of incidence, and its refractive index varies from 1.6585 to 1.4865. The minimum value is called the extraordinary index.

If a crystal of the spar be examined it will be found that the solid angles at two opposite corners are contained between three obtuse plane angles. The direction making equal angles with the planes of these obtuse angles is called the *optic axis* of the crystal. Iceland spar has but one such axis, and is therefore called uniaxial. Any plane normal to a refracting face of the crystal and parallel to the optic axis is called a *principal plane.* If the two solid angles bounded by the three obtuse angles are cut away by planes perpendicular to the optic axis, a beam incident normally on either of these surfaces passes through in the direction of the optic axis without bifurcation.

If the plane of incidence is perpendicular to the optic axis, both rays follow the ordinary law of refraction, and the index of refraction for the extraordinary ray is then 1.4865 for sodium light.

A prism may be cut from Iceland spar with its refracting edge parallel to the optic axis. A ray traversing it in the direction of minimum deviation is perpendicular to the optic axis, and the two components have the maximum separation.

Crystals like Iceland spar, in which the extraordinary index is smaller than the ordinary, are called *negative* uniaxial crystals. Quartz also produces double refraction; but in quartz the angular separation of the two rays is smaller than in the spar, and the indices of refraction are for sodium light 1.544 for the ordinary ray, and 1.553 for the extraordinary. Such crystals, in which the index of the ordinary ray is the smaller of the two, are called *positive* uniaxial crystals.

226. Theory of Double Refraction (T., 207; P., 247; B., 513; A. and B., 475; L., 286; D., 511). — Since the refractive index is inversely as the speed of light in any medium, it follows that for doubly refracting substances there are two speeds of light. For the ordinary ray the

speed is the same in every direction, while for the extraordinary ray the speed varies with the direction of the
ray between limits proportional to 1.486 and 1.658. It
has been found for Iceland spar that the phenomena can
be completely represented by supposing that a disturbance,
started in the interior of a crystal, gives rise to two concentric wave surfaces, one spherical and the other a flattened ellipsoid, with its polar diameter parallel to the optic
axis, and equal to the diameter of the sphere. The polar
and equatorial diameters are as 1.486 to 1.658, or as 0.603
to 0.676. The two wave surfaces therefore touch at the
extremities of the polar diameter, and the spherical surface
of the ordinary ray lies wholly within the ellipsoid of
revolution. A crystal in which the wave surfaces are thus
related is a *negative uniaxial* crystal. If the sphere en-
closes the ellipsoid, the two touching at the poles of the
ellipsoid, the crystal is a *positive uniaxial* crystal.

From these two wave surfaces the path of the two rays
may be determined by the method already employed in
Art. 187. Thus let
ic (Fig. 151) be the
direction of the in-
cident ray per-
pendicular to the
optic axis ab, and
let the plane of in-
cidence be a prin-
cipal plane. With
c as a centre, and

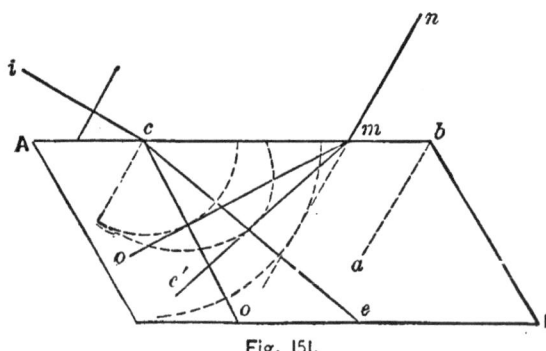

Fig. 151.

with ic produced as the major axis, draw a circle and an
ellipse with major and minor axes proportional to the ordinary and extraordinary indices of refraction, the radius of
the circle being the semi-minor axis. Draw also a larger

circle with a radius equal to the distance light will travel in air while the ordinary ray travels over the radius of the small circle in the spar. Then if *nm* is tangent to the outer circle at the point of intersection with *ic* produced, it is parallel to the incident wave; draw tangents from *m* to the inner circle and to the ellipse; the lines *co*, *ce*, connecting *c* with the points of tangency, give the direction of the ordinary and extraordinary rays respectively.

It is evident that if the optic axis were perpendicular to the plane of the paper through *c*, the section of the ellipsoid made by the plane of incidence would also be a circle, the extraordinary ray would lie in the plane of incidence for all angles of incidence, and would have its least refractive index and its greatest speed of transmission.

227. Polarization by Double Refraction (A. and B., 476; D., 513; B., 517). — The examination of light transmitted by Iceland spar, either by another crystal or by a piece of unsilvered plate glass, exhibits a marked difference between it and ordinary light. Let the extraordinary ray be intercepted by a screen, and let the ordinary ray fall on the plate glass at an angle of incidence of 57°. If now the plane of incidence coincides with the principal plane of the spar, the light will be reflected like ordinary light; but if the mirror is rotated about the beam of light as an axis the reflected light will grow dimmer and dimmer; and when the plane of incidence is at right angles to the principal plane of the spar, the light will fail altogether. If the rotation is continued, the light gradually regains its maximum intensity at 180°, and again fails at 270°. The extraordinary ray exhibits the same peculiarities in the same order, but it has its maximum brightness at 90° and 270°, and fails at 0°

and 180°. Both rays are therefore plane polarized, the ordinary in the plane of the principal section, and the extraordinary in a plane at right angles thereto. The vibrations composing the ordinary ray are considered to be at right angles to the optic axis, while those of the extraordinary ray are in a plane containing the optic axis and the incident ray, and may make any angle with the optic axis from 0° to 90°.

The plane of incidence in which the light is most freely reflected is called the *plane of polarization*. The vibrations composing the reflected ray are parallel to the reflecting surface.

228. Polarization by Reflection (T., 213; A. and B., 480; P., 234). — When light has been reflected from such surfaces as water, glass, polished wood, etc., at a definite angle depending upon the nature of each substance, it is found to possess all the properties of light polarized by Iceland spar or tourmaline. For glass there is a particular angle of incidence at which the reflected light is completely polarized, and this is called the *angle of polarization*. Brewster discovered that for this angle the reflected and refracted rays are at right angles to each other. The tangent of the angle of incidence then equals the index of refraction. For if the reflected and refracted rays are at right angles the corresponding angles are complementary, or

$$\sin r = \cos i.$$

Therefore $$\mu = \frac{\sin i}{\sin r} = \frac{\sin i}{\cos i} = \tan i.$$

It must not be inferred, however, that for every substance there is an angle of complete polarization. The polarization always increases with the angle of incidence up to a maximum and then decreases again, after passing the angle of maximum polarization. This maximum is called the *polarizing angle* of the substance. Only a few substances, with a refractive index of about 1.46, polarize light completely by reflection. If the substance is transparent the refracted ray is also polarized, and in a plane perpendicular to that of the reflected ray. The plane of polarization for the latter is the plane of incidence, and its vibrations are parallel to the reflecting surface.

229. Nicol's Prism (A. and B., 481; P., 254; S., 22). — A single beam of plane polarized light may be produced by transmission through a bundle of parallel plates at an angle of incidence of about 57°, the polarizing angle for glass; also by the passage through a plate of tourmaline cut parallel to its optic axis. Tourmaline is a doubly re-

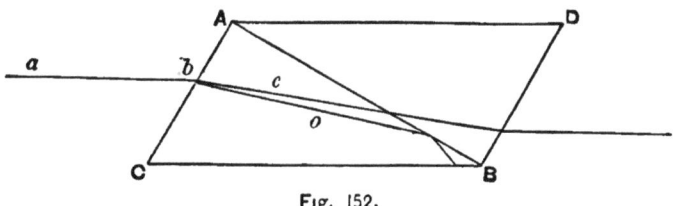

Fig. 152.

fracting substance and has the property of rapidly absorbing the ordinary ray, so that a plate 1 or 2 mm. thick is impervious to it. The extraordinary ray, on the contrary, it transmits. Tourmaline is not, however, a very transparent material, and the most effective arrangement for securing a beam of plane polarized light is a Nicol's prism. It is constructed of Iceland spar in such a way that one of the refracted rays is stopped by total internal reflection. A long rhomb of the spar has its terminal faces cut off obliquely so that the angle *ACB* (Fig. 152) is 68°. The plane of the figure is a principal plane. The rhomb is then cut through by a plane, the trace of which is *AB*, making the angle *ABC* 22°. The two faces of the section are polished and cemented together with Canada balsam, which has an index of refraction intermediate between those of the ordinary and the extraordinary rays.

When therefore a ray of light *ab* enters the prism it is divided into two rays, *bo* the ordinary, and *be* the extraordinary. But *bo* meets the Canada balsam at an angle somewhat greater than 68°, while the critical angle for the ordinary ray is 67° 31'. The relative index of refraction for the ordinary ray from Canada balsam to Iceland spar is $\frac{1.658}{1.532}$ (Art. 187), 1.532 being the index of the balsam from air. But the sine of the critical angle is the reciprocal of the relative index, or $\frac{1.532}{1.658} = 0.924$. This is the sine of 67° 31'.

Therefore the ordinary ray suffers total internal reflection in a Nicol's prism. The critical angle for the extraordinary ray from Canada balsam to Iceland spar, for this angle of incidence, is greater than 68°. It is not reflected at the first surface of the balsam, because it goes from a medium of lower refractive index for it to one of higher; and it is not reflected at the second surface of the balsam because the angle of incidence is less than the critical angle for the two media. Moreover, since the cemented section is at right angles to a principal plane of the crystal, the vibrations parallel to this section, and therefore those readily reflected, constitute the ordinary ray. Thus the extraordinary ray alone passes through.

The direction of vibration for the transmitted ray is the shorter diagonal of the end of the prism. This is in a principal plane. A Nicol's prism thus permits only those vibrations to traverse it which are in its principal plane, while it is completely opaque to vibrations at right angles to its principal plane.

230. Extinction of Light by two Crossed Nicol's Prisms. — When the light which has passed through one Nicol's prism falls upon a second, the amount transmitted will depend upon the relation of the principal planes of the two. The first prism is called the *polarizer*, and the second the *analyzer*. If their shorter diagonals are parallel, then the plane polarized light from the polarizer will compose the extraordinary ray in the analyzer, and will pass on through unaffected. But if the analyzer be turned around the beam of light as an axis, the transmitted beam will decrease in brightness, and will disappear entirely when the rotation has reached 90°. The Nicol's prisms are then said to be " crossed " ; the light from the polarizer now forms the ordinary ray for the analyzer, and is lost by internal reflection. In intermediate positions the rectilinear vibrations of the plane polarized extraordinary ray are resolved into two rectangular components in directions corresponding with the two planes of vibration in the analyzer. This resolution takes place in accordance with the usual mechanical law for the resolution of a motion into two rectangular components.

The intensity of the transmitted light is proportional to the $\cos^2 a$, a being the angle through which the analyzer has been rotated from the position of parallelism with the polarizer.

231. Effect of Interposing a Doubly Refracting Plate (T., 227; A. and B., 483; B., 528; L., 316; S., 28). — If two Nicol's prisms be placed with their principal planes

crossed no light will pass through them. Suppose now a
thin doubly refracting plate of mica or selenite to be in-
serted between them. It will be found that there are two
positions of the plate at right angles to each other in which
the field will remain dark. In all other positions of the
mica plate the light will be restored, reaching its maximum
intensity when the plate has turned round in its own plane
45° from the positions of no effect.

The mica or selenite is a doubly refracting substance;
and when the plate is in a position such that its two dirce-
tions of vibration coincide with the principal planes of
the polarizer and analyzer, the extraordinary ray from the
polarizer passes through without resolution into two components and is stopped by the analyzer. But in any other
position of the plate the case is different. Within it the
rectilinear vibrations of the plane polarized ray will be
divided into two components. Let Ox and Oy (Fig. 153)

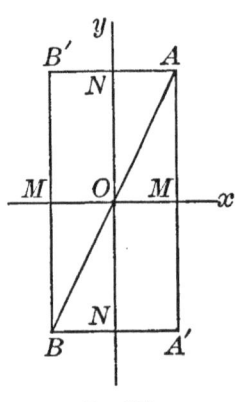

Fig 1S3.

be the two directions of vibration for the
mica or selenite plate. Then if OA represent the amplitude of vibration for the
incident ray, it will be resolved by the
plate into two vibrations whose amplitudes
are OM and ON. Both of these rays
will pass through the plate unchanged
except that one travels faster than the
other; a difference of phase will therefore result, dependent upon the thickness of the plate. The recombination of
these two vibrations on emergence will
result in elliptic vibration if the two beams produced by
the mica are not separated so far that they do not overlap.
But wherever the same ether elements are disturbed by
the vibrations of both the rays they will describe elliptical

orbits, and all the possible ellipses will be inscribed within the rectangle $AA'BB'$ (Art. 131). The amplitudes of the motion parallel to Ox and Oy are not altered, but the maximum displacement in one direction is no longer simultaneous with that in the other, unless the difference of phase becomes some multiple of 2π. The light emerging from the thin plate is then in general *elliptically polarized*. If OM equals ON and the difference of phase becomes an odd multiple of $\frac{1}{2}\pi$, the light will be *circularly polarized*.

OA must then be inclined at an angle of 45° with Ox, and the thickness of the plate must be such that one of the component vibrations suffers a relative retardation of one-quarter of a wave-length in traversing it; for the distinctive feature of double refraction is the difference of speed of the two rays.

Suppose now that the elliptically polarized light passes on to the analyzer. The introduction of the mica plate in general restores the light. If the mica plate is made to rotate in its own plane the light vanishes for successive positions differing by a quadrant of rotation. In these positions the directions of vibration for the interposed crystal coincide with the principal planes of the Nicol's prisms; and the light from the first prism passes unchanged through the crystal and is extinguished by the second prism. Midway between these positions of extinction the light transmitted by the system is brightest. If the mica is of such thickness as to produce circular polarization, the rotation of the analyzer does not alter the brightness of the transmitted light. When the elliptically polarized ray enters the analyzer, each of the two rectilinear components of the elliptical motion is resolved in the two directions of vibration for the analyzer. One pair of these components unite to form the ordinary ray, which is extinguished; the other pair form the extraordinary ray, which is transmitted.

232. Colors produced by Polarized Light (P., 244; S., 37; L., 319). — Colors produced from white light by means of polarization are due to destructive interference. But interference cannot take place between rays whose vibrations are at right angles to each other. The office of the analyzer is to bring together into one plane one component from each pair into which it resolves the two rays from the doubly refracting plate. Two rays of light polarized at right angles do not interfere like two rays of ordinary light.

Of the two pairs of components of the elliptically polarized light entering the second Nicol's prism, the one forming the extraordinary ray exhibits interference if the two components have opposite phases. The other pair then have the same phase; and if the analyzer is rotated through 90°, they form the extraordinary ray and are transmitted. If now the light is not homogeneous, then as the difference in phase depends upon wave-length, the retardation of one component compared with the other is such as to produce complete interference for a definite wave-length and partial interference for waves of approximately the same length. The corresponding colors then suffer complete or partial extinction, while the remaining colors of the incident light are transmitted, forming a colored beam.

233. Complementary Colors in the Two Standard Positions of the Analyzer (A. and B., 485; S., 35; L., 320). — When the principal plane of the lamina of selenite (crystallized sulphate of calcium) forms an angle of 45°

with the plane of vibration of both polarizer and analyzer, the most brilliant colors are obtained with a thin film. If now the analyzer be turned through 90° into parallelism with the polarizer, complementary colors of nearly equal brilliancy will appear.

If the two components resolved along Ox of the analyzer (Fig. 153) annul each other, the corresponding color is wanting in the light vibrating in this direction; but at the same time the components along Oy are added, and the same color is found undiminished in the light whose vibrations are confined to this plane. For other colors "the relative retardation is different; but for each vibration period, the component in the direction Ox combined with that in the direction Oy represents the total light for that period in the beam entering the analyzer." Hence the sum of the two represents all the light entering the analyzer; and therefore the light transmitted when the Nicol's prisms are crossed must be complementary to that passing when they are parallel, if the incident light is white.

Thick plates of selenite do not produce color. If the doubly refracting plate is thick enough to produce a relative retardation of several wave-lengths for extreme violet, it will produce a retardation of half as many wave-lengths for red, and an intermediate number for intermediate colors. Hence with crossed prisms extinction of these colors will take place. These losses will be distributed at about equal distances along the spectrum. But the transmitted light will consist of the different colors in nearly the same proportion as in white light, and it will therefore be white, but of diminished intensity.

If two plates of selenite, of exactly the same thickness, and therefore producing the same tint, are superposed in such a manner that their principal planes coincide, or so that the extraordinary ray through the one is also the extraordinary ray through the other, they exhibit another color, which is precisely the same as that produced by a plate of double the thickness of either. But if they are superposed so that their principal planes are perpendicular to each other they produce no effect. The screen remains dark. The ray which travels the more slowly in the first lamina travels the more rapidly in the second, and the two emerge together as if a plate of annealed glass had been interposed. The two rays which leave the compound plate have no difference of phase, and cannot exhibit interference and color. If the crossed laminæ are of unequal thickness, the effect is the same as that produced by a single lamina whose thickness is their difference.

234. **Colored Rings produced by a Plate cut at Right Angles to the Optic Axis (L., 326; S., 95).** — Extremely beautiful and interesting phenomena are produced by plates of uniaxial crystals *cut perpendicular to the optic axis*, such as a section of Iceland spar, in a beam of *converging* plane polarized light. The central ray of the converging cone should be normal to the plate. It then passes through the crystal along the optic axis without undergoing double refraction. But all other rays in the cone traverse the crystal section obliquely and are doubly refracted. The further the ray is from the axis of the cone the greater is the obliquity of its path through the plate and the greater the thickness traversed. The retardation of the one component behind the other is also greater; and since at equal distances from the optic axis both causes determining difference of path of the two doubly refracted rays are equal, it follows that the same difference of path must exist for all points of a circle con-

ceived as drawn upon the screen around the intersection with it of the axial ray of the cone. Hence a system of concentric colors appears on the screen in iridescent rings like those of Newton's rings, obtained by pressing the convex side of a plano-convex lens against a plate of plane glass.

When the polarizer and analyzer are crossed the colored rings are traversed by a black cross. This is explained as follows : Since the optic axis is perpendicular to the surface of the plate of spar, every straight line drawn through the centre of the system of rings is the trace of a principal plane. The vibrations of the ordinary ray are perpendicular to a principal plane and therefore tangential to all the concentric circles ; those of the extraordinary ray are in a principal plane or radially in the circles. Hence in the two diameters, representing the planes of vibration of the analyzer and polarizer, the directions of vibration in the thin plate correspond with those of the polarizer and analyzer, and therefore in these two directions the thin plate produces no double refraction and has no effect on the light. In all other directions, the tangential and radial directions of vibration for the plate are inclined to those of the polarizer and analyzer, and therefore double refraction takes place with interference and colors.

If the analyzer is turned so as to be parallel with the polarizer a white cross takes the place of the black one. The reason is evident, the colored rings being then simply projected on a bright field as a background.

235. Double Refraction in Quartz (A. and B., 491 ; S., 41). — Quartz is a positive uniaxial crystal and gives an ordinary and an extraordinary ray. When a quartz plate, cut perpendicular to the optic axis, is interposed between the polarizer and analyzer, the effects differ greatly from those produced by Iceland spar or by selenite. Let the plane polarized light fall on the quartz normally. Then with crossed Nicol's prisms, the light is restored and is unchanged by the rotation of the quartz through any azimuth. Homogeneous light will be extinguished by the rotation of the analyzer through a certain angle, indicating that the effect of the quartz is to rotate the plane of polarization. The amount of the rotation is the same as the angle through which the analyzer must be turned to produce extinction. Some specimens of quartz rotate the plane of polarization to the right, in relation to the direction of the light, and are therefore called *right-handed.* Others rotate it to the left, and are called *left-handed.* The amount of rotation depends upon the thickness of the plate and the wave-length of the light. Hence with white light the effect of rotating the analyzer is to quench in succession the several colors as the plane of polarization for each is reached. The resulting colored beam changes its tint continuously as the analyzer rotates. For a given plate the angle of rotation of the plane of vibration varies nearly inversely as the square of the wave-length. For a quartz plate one mm. thick Brock found that the B line was rotated through 15° 18′, and the G line, 42° 12′.

In the explanation of these phenomena given by Fresnel, the vibrations of the two rays for quartz are supposed not to be rectilinear but circular, and in opposite directions. The light of each ray is circularly polarized. The circular motion of the ether is right-handed for one ray, and left-handed for the other. When these two motions are impressed upon the same portions of the ether at the same time, the result is plane polarized light (Art. 32). But one of these motions is transmitted with greater speed than the other; in other words, its period of motion in the circle is slightly less; and since the resulting simple harmonic motion is in the plane of symmetry, that plane rotates in the direction of the component motion of shorter period. The rotation of the plane of polarization of the resulting plane polarized light is in the same direction as that of the plane of symmetry.

Many liquids, including a solution of sugar, rotate the plane of polarization. An instrument for comparing this rotation with that produced by quartz is called a saccharimeter.

INDEX.